Offbeat Landmarks: Exploring Around Stavanger

Rosslyn Nicholson

Map data: Kartverket, Geovekst, and kommunes

ISBN: 978-82-303-3581-9

Disclaimer: Although effort has been made to ensure that the information in this book is correct at the time of publishing, the author can make no warranty about the accuracy of its content and cannot accept any responsibility for any consequences arising from the use of this book. You are responsible for you own safety while engaging in any activity outdoors. Just because an activity is described in this book, it does not mean that it will necessarily be safe for you. Be aware that trail and road conditions may change over time, and so may weather. Make allowances for the limits of your party.

Above: A "birds-eye" view of Stavanger from the summit of Lifjell

Introduction

What this book is and isn't:

This book is not a guide book. These already exist, and there seems little reason to add another book describing the most popular local attractions to the shelf. Instead, it's a collection of curiosities and oddities. It visits places that tourists don't often see, or looks at familiar sites with a new slant. Stavanger is a town rich in history, and traces of events long past are here for us to see today. There is so much here in Stavanger that this little collection is not even close to being a comprehensive history or geography of the region: those may be found in the local library.

The chapters in the book describe things that I noticed. A different person would have noticed different things, and written a different book entirely. That is really the point of the book: I hope that in finding out about some of my favourite places and stories, readers will be inspired to explore for themselves.

There are suggestions for excursions within each section. I have tried to avoid paid-for attractions as these tend to be well advertised and easier to find, although several of the local museums are included, as are one or two restaurants with historical connections. It is my hope that these will act as a starting point for many more excursions, leaving the best question of all: what will you discover?

Rosslyn Nicholson

Stavanger, May 2017

Below: An ancient stone cross, dated to the first half of the eleventh century: found in Tjelta and now outside Stavanger Museum. The meaning of the name Tjelta is not known. Like Sola nearby, it does not come from a language that we know today.

Contents

Chapter 1: Ancient Hills.....2
- 1.1 The New Cutting: A glimpse into the past.....4
- 1.2 The "Moon" Rock: A glimpse into the present.....10
- 1.3 The Collision: Discovering a meteor crater.....14
- 1.4 The Collapse: Changing the course of a river.....18

Chapter 2: It isn't all about Fjords.....24
- 2.1 Undeknuten and the lines on the rock.....26
- 2.2 Vassøy: The exotic and the erratic.....30
- 2.3 The Pebble Beaches.....35
- 2.4 Kettles for Giants.....42

Chapter 3: Of Monuments and Mysteries.....50
- 3.1 The Petroglyphs: Rock Carving in the Bronze Age.....52
- 3.2 Standing Stones and what becomes of them.....58
- 3.3 The Standing Crosses.....66

Chapter 4: Getting around the Jaeren.....72
- 4.1 Kongevegen: "The King's Road".....74
- 4.2 Building Roads: "The Western Highway".....80
- 4.3 The Wonderful Winding path from Vigdel to Rege.....86

Chapter 5: The Town of Wooden Houses.....90
- 5.1 Valbergtarnet: The Watch Tower.....92
- 5.2 Living the High Life, Merchant Style.....94
- 5.3 The Country Residences.....98
- 5.4 Gamle Stavanger: streets that were nearly demolished...104
- 5.5 Around Pedersgata.....108
- 5.6 The Villas in "Paradise".....114
- 5.7 The Gingerbread Town.....118

Chapter 6: Reinventing Retail Through Changing Times.....122
- 6.1 The River Figgjo: Manufacturing quality tableware.....124
- 6.2 Fretex: Upcycling History.....126
- 6.3 Creating Fun: Kvadrat and the City Centres.....128

Further Resources:.....132

Acknowledgements:.....134

Above: The Mån valley: this typical Norwegian scene shows just how much of the terrain consists of exposed rock.

Chapter 1:
Ancient Hills

Above: Patterns in the rock on the coast at Langøy

Ever since the North Sea oil boom turned the Stavanger area into the powerhouse of the Norwegian economy in the 1970s, the conurbation around Stavanger and Sandnes has been growing and thriving. As of 1 January 2016, it was the third largest urban area in Norway, behind Oslo and Bergen. Many people have moved into the area, and discovered that city life here is a little different. Stavanger is a city set amidst nature so beautiful that cruise ships bring flocks of tourists. By the side of the business parks lie lakes, hiking trails, and wilderness.

Everybody has a walking trail in their area, and each neighbourhood boasts at least one outside play area. The town marinas are crammed full with boats. Fresh air and outdoor life are very much part of the culture here. Newcomers are encouraged to invest in good rain gear, and join in. Few refuse to do so; the hills, mountains, beaches and lakes are too enticing.

Most of Norway is great for outdoor adventures, but is not land that is suitable for farming. The Jaeren, to the south of Stavanger and Sandnes, is farmed, but even there the fields had to be cleared of boulders before they could be ploughed. Outside the Jaeren, the land is generally rocky, and the soil thin. Forest, bog and moorland predominate, with a few farms in the valleys and some summer grazing clinging to the hillsides.

What makes for conditions like this? The answer has a great deal to do with the underlying rock. Norway is sometimes described as a land of rock: a place where stone is piled upon stone. This chapter is about the rocks and the stones.

Above: The Three Swords monument at Møllebukta, a popular beach not far along the shore from the rock cutting in this section.

1.1 The New Cutting: A glimpse into the past

The mountains of Norway attract less attention than the fjords. They are more rounded and less spectacular than many other mountain ranges, more bog and less alpine meadow. This is partly because these mountains are ancient. They were built about 400 million years ago, pushed up into a huge mountain chain by continental plates colliding. Over the eons, erosion and continental drift have taken their toll. Huge chunks of the land drifted away from Europe and are now to be found in North America. The original high, jagged peaks have been ground down by wind, weather, frost and ice. The top layers have been scraped away, so that the rocks we see now on the surface were once buried deep under tons of mountain.

Top: The new rock cutting.
Above: The new bridge on the path from Møllebukta to Madlasandnes. The rock cutting is just beyond the bridge.

They were buried so deep, under heat and pressure, that the rocks became partly molten and could flow a little. Minerals of different densities separated out into layers. The ability to flow allowed the rocks to be squeezed into waves and folds under the pressure of mountain building. At the same time, some mineral crystals melted into the mix and new, different minerals recrystallized back out again when later, the rock slowly cooled. Rocks that have been altered this way, deep underground, are called metamorphic rocks. An example seen around Rogaland, and in this example, is called gneiss (pronounced nice).

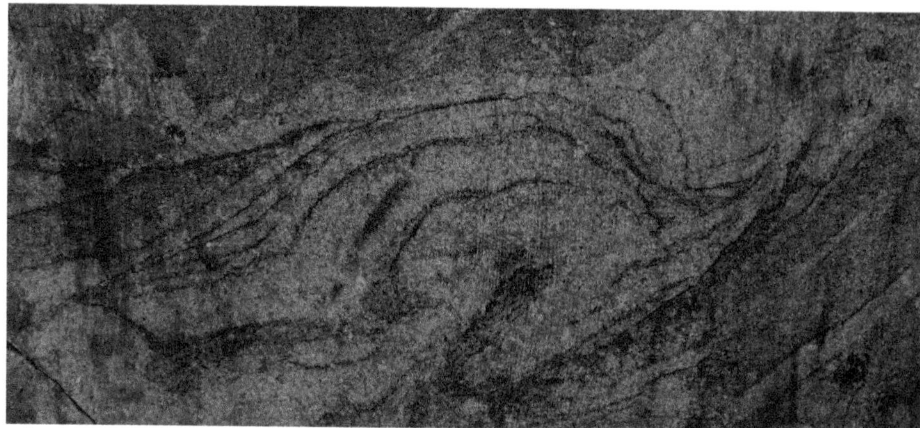

Above: The rocks at the new path cutting near Madlasandnes. When they were buried deep in the earth, under incredible heat and pressure, lighter and darker coloured minerals separated out into layers. Later, when the rock was still semi-molten, these layers were squeezed and folded under immense pressure as mountain chains were built above them.

At times, as this semi-molten rock was becoming gneiss and was cooling, or perhaps after it was completely solid, molten rock at very high temperatures was pushed up from below. Liquids can't be squeezed like gases. Instead, the huge hydraulic pressure forced the molten rock up into weak spots: joints or cracks in the rock above. As the hot molten rock filled the cracks in the gneiss, it cooled, forming sheets and ribbons called dykes inside the gneiss, or host rock.

Above: A dyke cuts through the "host rock" of gneiss at the new path cutting near Madlasandnes.

It is sometimes difficult to see dykes on the surfaces of

Fig. 1.1: Formation of gneiss and dykes

1

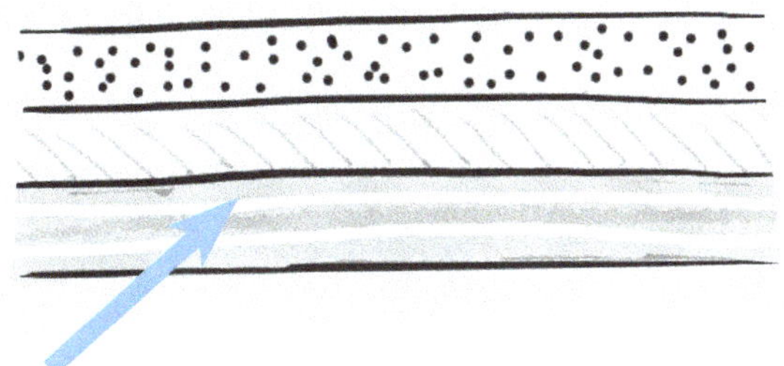

Rock buried at high temperature and pressure changes into gneiss

2

Pressure from shifting continental plates pushes rock up to form mountains and squeezes it into folds

3

Super-hot molten rock from a source deeper in the Earth's crust is forced through cracks in the upper rock layers and solidifies there as dykes

4

Over millions of years, the mountains are eroded away, leaving the gneiss and dykes at the surface.

outcropping rocks (see section 1.2). The rock surface is exposed to the weather, where frost and water slowly but surely change it. Minerals in the surface rock become altered, and it is difficult to tell what minerals lie beneath this crust of "weathered" rock without wielding a geologist's hammer. Road cuttings, where the rock faces are freshly cut and haven't been weathered, are often a good place to observe the rocks. In 2015, a new path cutting was made along the side of Hafrsfjord. Here is a chance to enjoy looking at an example of dykes without the worry of cars speeding past.

Above: The thin layer of contact metamorphism between the dyke and the gneiss

The cutting has been blasted through a mass of gray and pink banded gneiss. The gneiss is criss-crossed with bands of a different, dull grey coloured rock. These bands were originally the molten rock that was injected into the gneiss so long ago. Just at the edge of the dykes, where they come in to contact with the gneiss, there is a thin layer in the gneiss that looks different. The gneiss was altered when the hot rock of the dyke, came into contact with it. This is called contact metamorphism.

Looking closely at one of the dykes, it is possible to see individual crystals within it. The crystals in the middle of the dyke are larger than the crystals at the edges (in fact, the crystals near the edges of the dykes are so small they are very difficult to see without a magnifying glass). This is because the molten rock that came into contact with the colder, solid rock, cooled more quickly than the molten rock in the centre of the dyke that didn't come into contact with the colder rock. This gave the crystals at the centre of the dyke more time to grow, so they grew bigger than the crystals at the edge.

Above: The rock near the centre of the dyke cooled more slowly, so the crystals there had more time to grow, and are larger.

The path cutting was first made in 2015. At the time of writing, in 2017, the surface was already beginning to change because of weathering, Fortunately, the complete process of weathering will take many years, and it should be possile to enjoy viewing these rocks for several years to come.

Excursion 1.1:

Walk from Møllebukta to Hestnes

A popular walk along the side of Hafrsfjord, this trail is often more sheltered than others during stormy weather. The rock cutting is best viewed when the rock is wet, but be aware that the cutting is also a little slippery when wet!

Access: car, although there are several bus stops quite close

Length: about 9km there and back, but very easy to shorten

Climb: negligible

Grading: easy, suitable for pushchairs and wheelchairs

Facilities: toilet and ice cream kiosk (summer only) at the car park at Møllebukta.

To reach the start of the walk:
From Stavanger, take Rv 509 (Madlaveien) towards Tananger. After about 3.6km, turn left towards Sola. The car park is signposted on the right hand side of the road after about 1km.

Walk Directions:
From the car park, stand facing the water and take the gravel path from the right hand side of the beach. You will go past a bronze statue.
Continue walking, with the fjord on your left hand side. The trail is very well signposted. The path cutting, and the new bridge by its side, are located just before the coastline turns north at Madlasandnes.
The peninsula at Hestnes is often a good place to turn around and return, but the path along Hafrsfjord continues all the way to Hafrsfjord Bridge and beyond. Either return back along the track whenever you are ready, or return to Møllebukta along the roads for a circular walk.

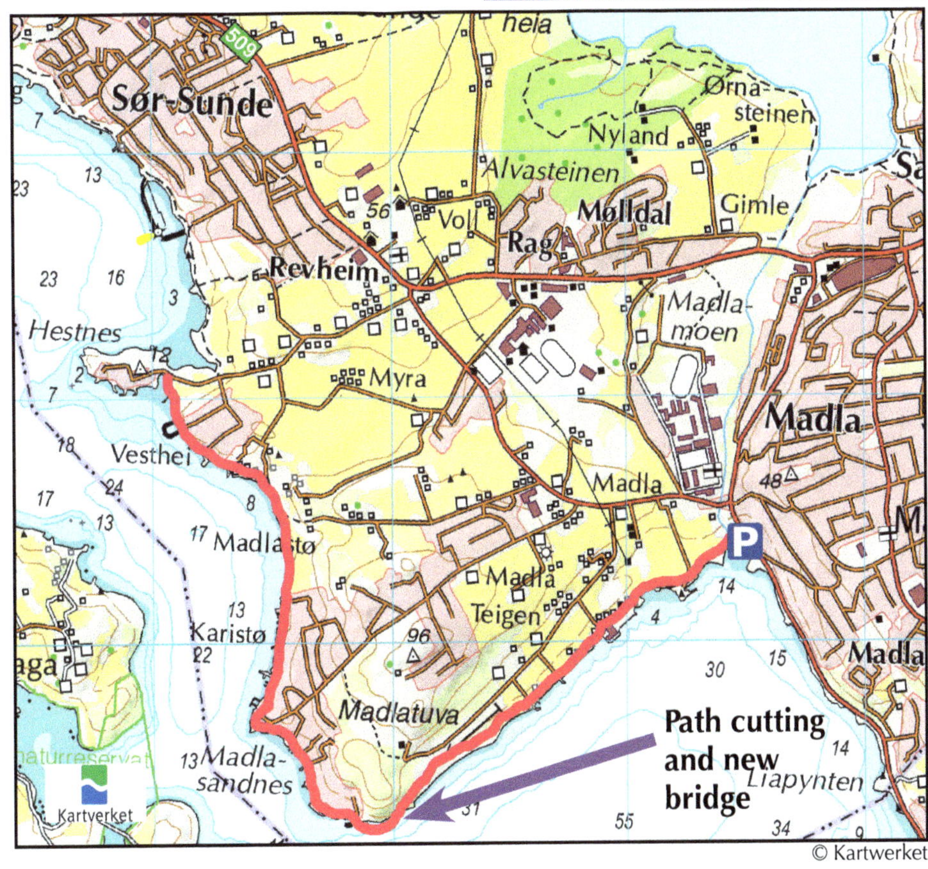

© Kartwerket

Above: Eigerøy Fyr

1.2 The "Moon" Rock: A glimpse into the present

Top: Glacial striae near Eigerøy Fyr (see Section 2.1)
Above: A large dyke cuts through this knoll near Eigerøy Fyr

Eigerøy Fyr is one of many lighthouses on the coast of Norway. Eigerøy lies to the west of Egersund, where the town gives way to beaches and summerhouses. The lighthouse is to the far west of the island, beyond a World War II lookout post. It was built in 1854 and, ever since it was automated in 1989, shares the island only with sheep. Eigerøy Fyr is a majestic sight. It towers above the smooth rock of the island shore, 32.9 meters high and is the oldest cast iron lighthouse in Norway.

The rock beneath the lighthouse is special too. Moon rock is a rare and precious treasure on Earth. Between 1969 and 1972, 6 of the Apollo missions retrieved a total of 382 kilograms of rock, core samples, pebbles, sand and dust from the moon. In addition, 3 remote controlled Soviet

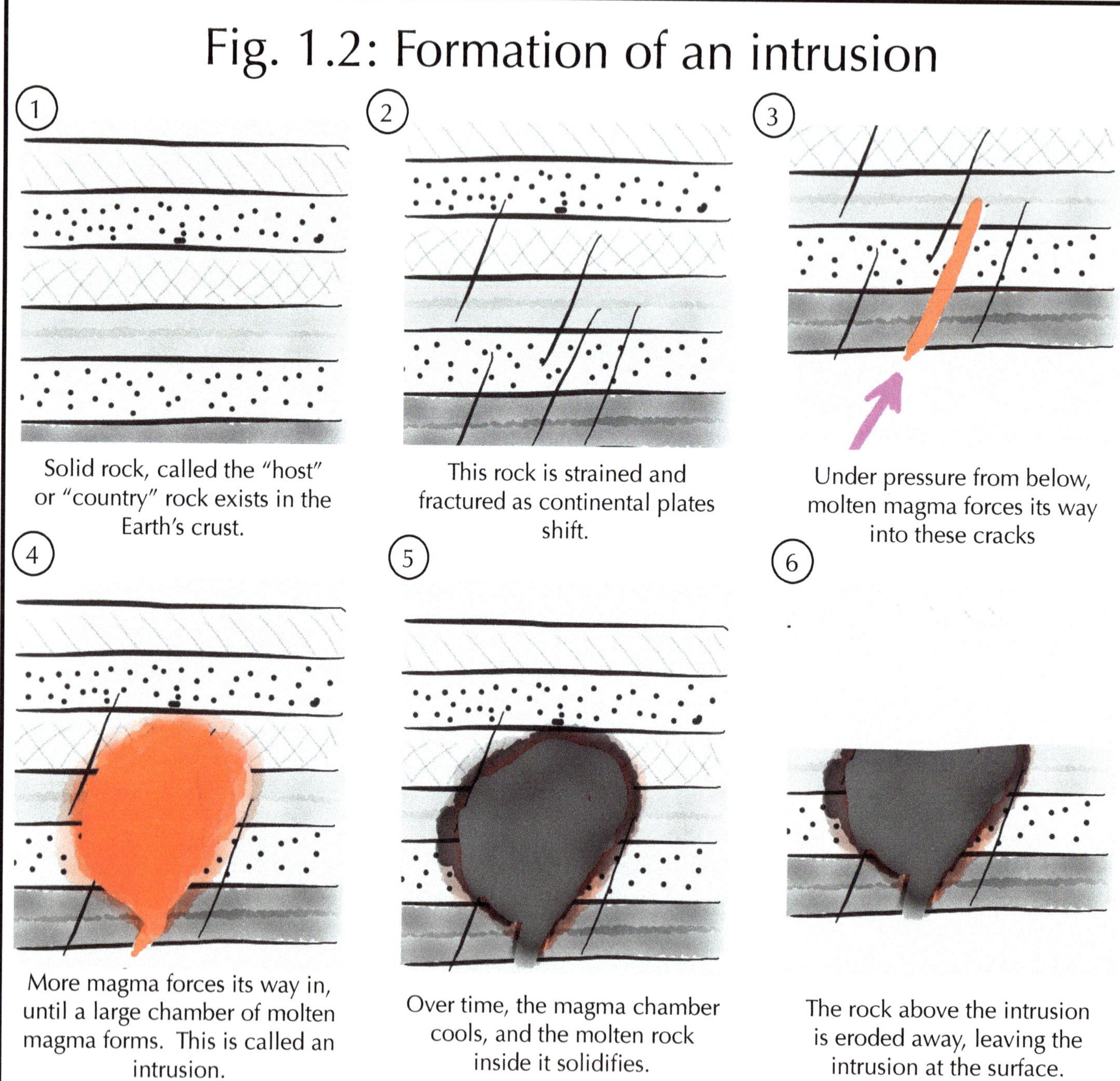
Fig. 1.2: Formation of an intrusion
1
Solid rock, called the "host" or "country" rock exists in the Earth's crust.
2
This rock is strained and fractured as continental plates shift.
3
Under pressure from below, molten magma forces its way into these cracks
4
More magma forces its way in, until a large chamber of molten magma forms. This is called an intrusion.
5
Over time, the magma chamber cools, and the molten rock inside it solidifies.
6
The rock above the intrusion is eroded away, leaving the intrusion at the surface.

Above: A lookout post from World War II near Eigerøy Fyr

aircraft collected 300g of moon rock and carried it back to Earth. Some samples were distributed around the world as goodwill gifts. Several of these samples have since gone missing, and some have reappeared in auction houses and private collections amidst tales of theft, misappropriation, and desperate sales that rival any famous gemstone. Most of the rest of the Apollo moon rock samples are carefully curated in the Johnson Space Centre in Houston. They are lent out for study and teaching purposes.

Geologists studying the moon rock samples found that there were two main types of rock amongst the samples. The first is a kind of basalt, thought to have come from lava flows that erupted early in the life of the moon as the planet was still cooling. The theory is that the impact of meteorites caused cracks in the surface rocks, through which the molten lava could flow. The basalt flows are found in the lower areas of the Moon, and cover about 26% of the side facing the Earth, but only 2% of the side facing away.

The second kind of rock collected from the surface of the moon is called anorthosite. Anorthosite is a light-coloured rock, which is rich in a mineral called plagioclase feldspar. It is believed that the anorthosite formed on the surface of the moon when the planet was still molten: the plagioclase feldspar crystallised out of the molten mixture and, because it was less dense, floated to the top. More dense minerals sank towards the core of the moon.

The good news is that if you aren't able to buy, borrow or steal a sample of anorthosite from the moon, this rock is visible in some places on Earth too. One of these places is Eigerøy Fyr. It is thought that the anorthosite at Eigerøy Fyr formed part of a huge chamber of magma (molten rock), which cooled about 20 kilometres beneath the surface of the earth some 930 million years ago. After cooling, it was uplifted during an episode of mountain-building, and now that the mountains that were on top of it have been eroded away, it is left at the surface.

Excursion 1.2:

The Geological Trail at Eigerøy Fyr

The walk to Eigerøy Fyr is a short and pleasant outing. There is a geological trail covering the first part of the walk. A guide to the trail, with excellent explanations, can be downloaded (see below).

Access: car

Length: about 4.5km to Eigerøy Fyr and back; the geological trail is shorter.

Climb: negligible

Grading: easy

Facilities: none, but sometimes a Sunday cafe at Eigerøy Fyr

To reach the start of the walk:

By car: From Stavanger, take the E39 south towards Kristiansand. About 15km south of Vikeså, turn right along Rv42 towards Egersund. After about 10km, just outside Egersund, take Rv502 towards Eigerøy. After 1.5km, cross a bridge and turn right immediately afterwards. After just over 5km, the road to the car park is signposted to the right. The car park is at the end of the road.

Walk Directions:

NOTE: The excellent guide to the Geology Trail is availalble for download from the Magma Geopark website (www.magmageopark.no). Don't forget to download the brochure before you begin, because although there is a general information board at the car park, at the time of writing there were no information boards along the trail itself.

From the car park, cross over the narrow bridge and follow the track through the field. The geological guide will show you the points of interest along the way. After the end of the geological section of the trail, the path to the lighthouse is signposted and marked with red dots. There are plenty of places along the way to stop, play, picnic and enjoy the view.

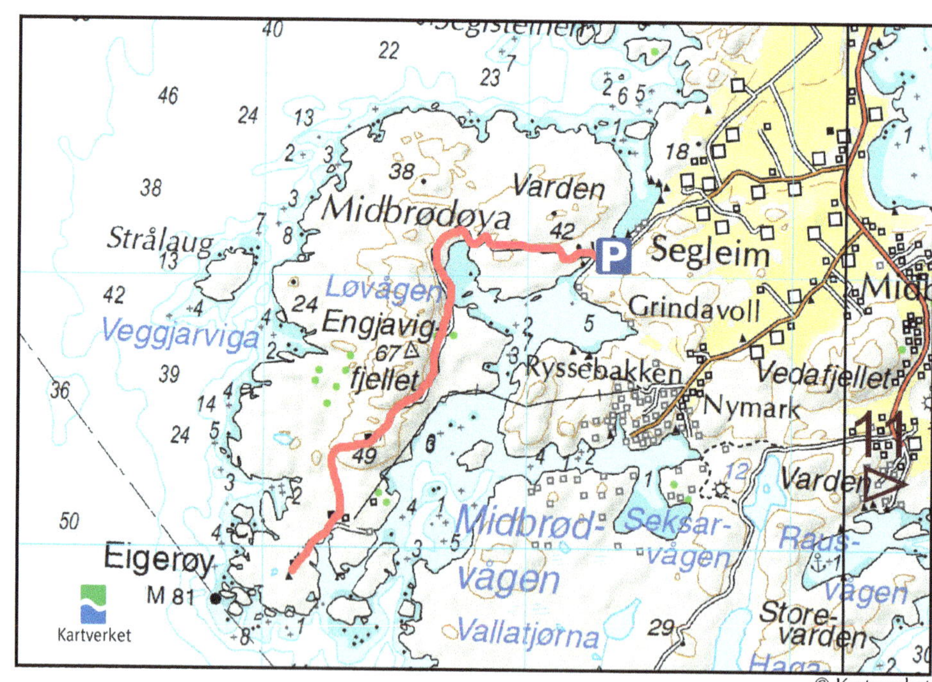

© Kartwerket

1.3 The Collision: Discovering a meteor crater

Our solar system contains smaller rocks than planets, mostly orbiting the sun in the Asteroid Belt. The rocks from the moon in the Kennedy Space Center were carefully collected and brought back to Earth with the utmost care. Space rock can arrive on Earth more violently: literally directly from the sky. When a lump of rock, or meteorite, gets trapped in Earth's gravity field, it falls towards the surface. Most of the time, the intense heat of entering the atmosphere at incredibly high velocity burns the meteorite before it reaches the ground, and we see a shooting star. Occasionally, the meteorite doesn't burn completely, and hits the ground.

Depending on the size of the meteorite, the force can be bigger than a bomb blast, and blow a crater miles wide. Meteorites have collided with the moon, too.

Top: The viewing area for the Ritland Crater
Above: Looking south from the crater viewpoint towards the two lakes at the bottom of the crater

On the moon, these craters remain mostly unchanged over time and can be seen from the Earth. On Earth, erosion, deposition of sediment, and mountain building obscures the craters. There are huge, famous meteorite craters in Arizona, Australia, Mexico, Russia and other places. There is also a rather smaller crater just up the road from Stavanger, near Hjelmeland.

Above: Kleivalandsåna near the Ritland crater car park

In 1985, a geologist called Fridtjof Riis visited the Ritland area for the first time. While there, he noticed unusual rock formations that couldn't be explained by "normal" geological process. In 2001, he saw the first known meteorite crater in Norway: Gardnos crater in Hallingdal. He noticed the similarity between Gardnos and what he had seen in Ritland, and wondered if the rocks in Ritland might be part of an impact crater. He contacted the University of Oslo and, with the help of a research grant, the geology of Ritland was studied in depth.

The first discovery was of a rock type called suevite. Suevite is formed under the immense pressure of an impact, but there are other rock types that can look just like it, so it didn't prove that there was an impact crater in Ritland. The breakthrough came under the microscope, with the discovery of a particular crystal form of quartz called "shocked quartz". The only known natural way to create shocked quartz is via a meteorite impact. This proved that Ritland is indeed an impact crater. The story of the meteorite impact was then pieced together from the remaining rocks at the site.

The meteorite hit the earth about 500 million years ago, landing in what was then a shallow sea with clay at the bottom. Although the meteor itself is thought to have been between 100 and 150 meters wide, it evaporated upon impact, leaving a crater about 2.7 kilometers wide and 400 meters deep. The rocks in the area were shattered, with some of them being thrown over 4 kilometers away from the impact site. Some rock at the point of impact was melted, and formed the suevite found here.

After the crater was formed, stones and pebbles tumbled over its rim and landed at the bottom of the hole. Then, the crater was filled with sea sediment. Fossils, including trilobites, have been found in this sediment. After the crater had been filled, about 200 million years ago, shifting continental plates built a huge mountain chain here. The crater was buried several kilometers under the mountains, and the rocks were changed under the heat and pressure. The clays became shale, sand became sandstone, and the shattered rocks at the bottom of the crater were also altered.

Time eroded away most of the mountains and parts of the crater site. The western crater rim and most of the shale that had filled it is now gone, but the eastern rim of the crater is still on view for us to study and admire.

Right: The mountain farm at Trodlå-Tysdal may be reached in a day hike from the Ritland crater car park. For more information, see www.trodla-tysdal.no

Excursion 1.3:

Visit the Ritland Crater Viewing Area

An easy hike leads to the viewing area which, even without the crater, would be spectacular.

Access: car	
Length: about 6km	
Climb: about 300m	
Grading: moderate: muddy in places	
Facilities: none	

To reach the start of the hike:

By car: From Stavanger, take the Tau ferry. After leaving the ferry, take the Rv 13 to Hjelmeland (about 45km). At Hjelmeland, take Rv660 (there is a sign for the Ritland crater). After about 16km, look for a road on the right, signposted Kleiveland. Take this turning (also signposted to the Ritland crater), and the car park is at the end of the tarmac road on the right hand side.

Hike Directions:

From the car park, take the track signposted to the Ritland crater viewing area. Follow the trail marked with red T-markers up past a farm and on to a track.

Continue on the track, following the markers uphill. When you reach a path junction, take the left track, signposted to the Ritland Crater viewpoint. From here, the path is not marked with T markers, but is reasonably straightforward, with signposts in crucial places!

At the time of writing, there was one fork in the track that was not signposted (marked with an arrow on the sketch map). Take the right fork here!

From the viewing area, the trail is signposted back to the car park down to the marshy, but delightful valley floor.

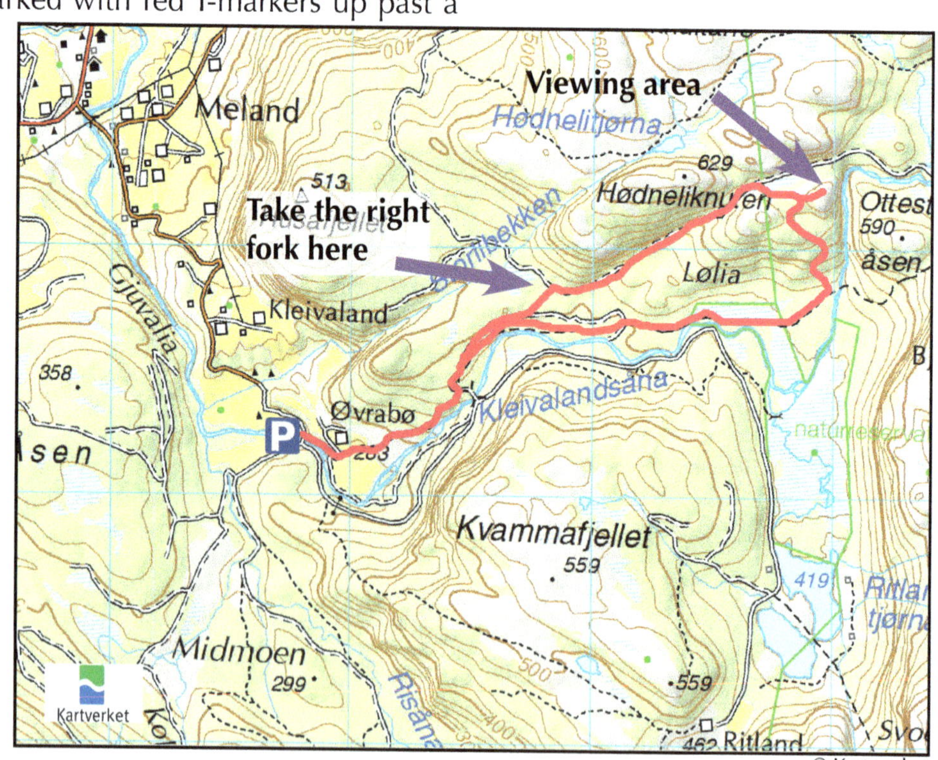

© Kartwerket

1.4 The Collapse: Changing the course of a river

Byrkjedal is a valley surrounded by steep cliffs left behind by glaciers. Early in the twentieth century there was a dairy here, to process milk from the sheep and mountain goats kept by local farmers. As Stavanger developed, the road through the valley became busier with traffic going to and from cabins in the mountains higher up. The dairy reinvented itself as a wayside stop, serving welcome drinks and delicious food. Nowadays, it is a thriving hotel, cafe and shop large enough to serve coach tours from the cruise ships that visit Stavanger. It's easy to grab refreshments and carry on driving, but just up the road is a rockfall claimed to be the largest in Europe.

Near the end of the last ice age, there was a glacier that remained just about stationary in Byrkjedal. A pile of glacial debris (stones

Top: Byrkjedalstunet
Above: Looking southwest from Gloppedalslura

and gravel) accumulated at the bottom of it. This is called "terminal moraine". The terminal moraine acted like a dam, so that when the glacier began to retreat, the meltwater from it couldn't flow down the valley towards Vikeså, because the terminal moraine was blocking the way. Instead, it had to form a shallow lake.

The glacier had retreated further up the valley now called Hunnedalen, and as it continued to melt, the water coming from it took with it sediment that washed into the shallow lake. The sediment built up to form terraces, which can still be seen near Byrkjedal today.

Sometime after the lake had formed, the rocks on the steep south side of the valley above the terminal moraine collapsed. This wasn't a little rockfall that caused a pile of scree to roll down the slope. It was a rockfall on a much bigger scale. The whole cliff side fell down at once. Blocks of granite as large as houses tumbled down the cliff and landed on top of the terminal moraine that was damming the

Above: Looking northeast from Gloppedalslura

Fig. 1.4: Formation of the Rockfall

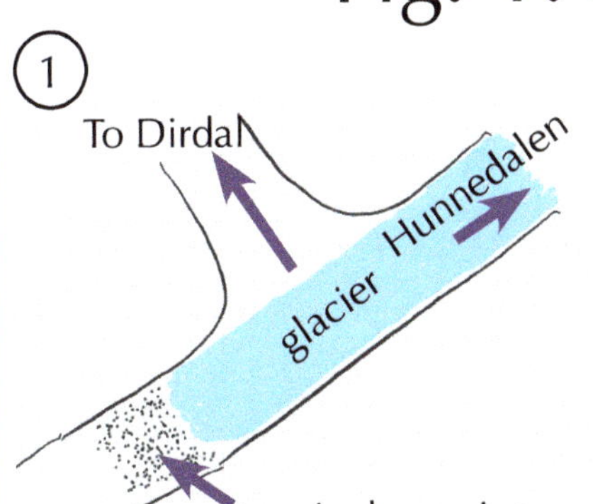

A glacier stays stationary near where Byrkjedal is today. Gravel and stones build up at the bottom of it, forming a "terminal moraine"

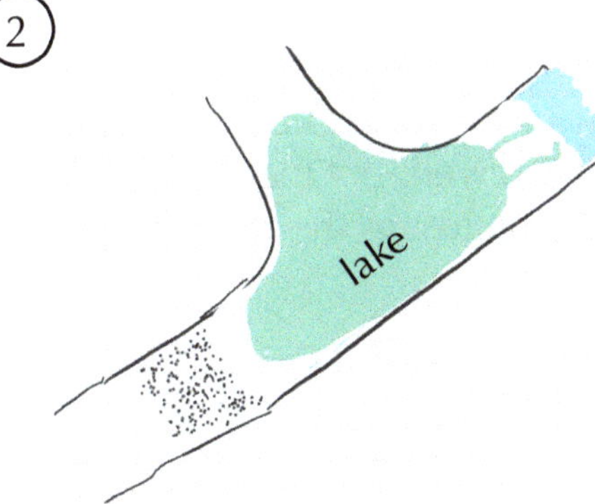

As the glacier retreats, the terminal moraine acts as a dam, stopping the meltwater from draining away. A lake forms behind the terminal moraine.

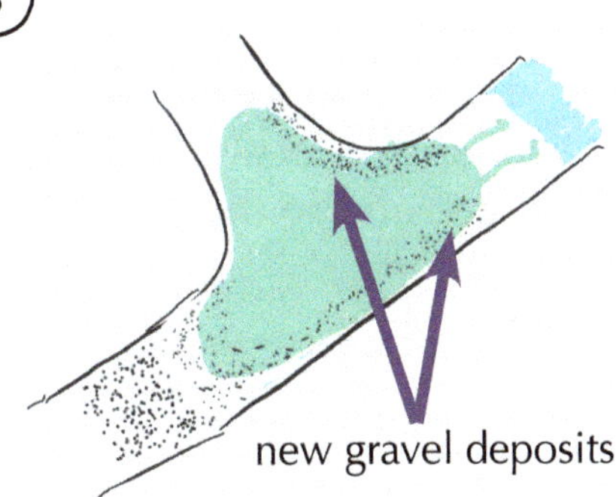

Meltwater from the retreating glacier carries sand and gravel with it, which is deposited in the lake as sediment.

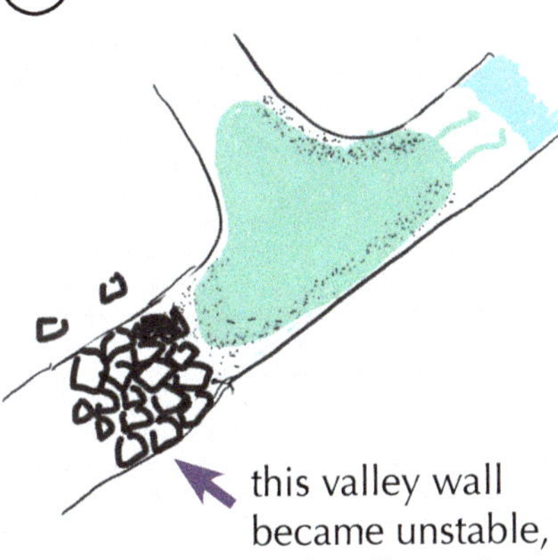

The valley wall to the south of the terminal moraine collapsed in a huge rockfall. Some rocks were thrown up against the far side of the valley.

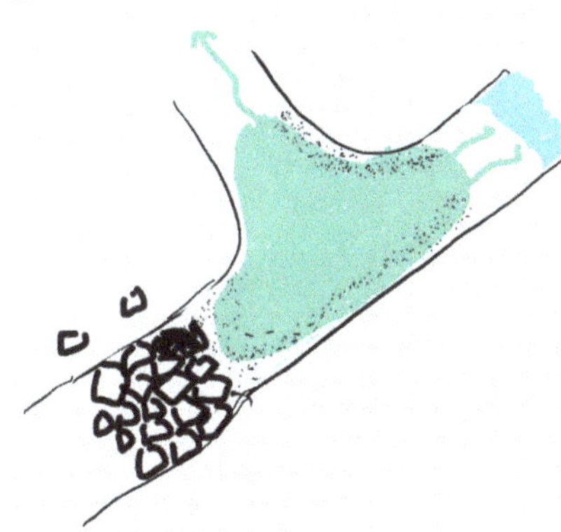

With the valley to the southwest now blocked by the terminal moraine with the rockfall on top of it, the lake is forced to drain towards Dirdal.

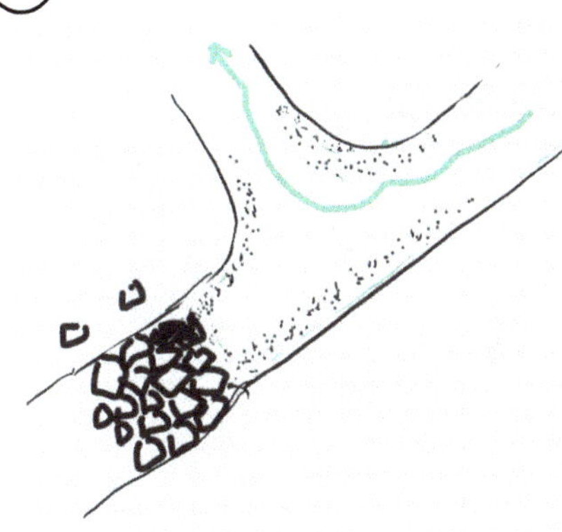

Today, the lake has drained away, but the gravel deposits remain at the side of the valley, and the river still flows towards Dirdal.

lake. The force of the rockfall was so great that some of the boulders were thrown up to the north side of the valley. This huge pile of scree today is called Gloppedalslura. A narrow road winds up and over it, making it easy to visit and just a five minute detour from Byrkjedal.

During the Second World War, the Resistance used Gloppedalslura as a hideout. The locals knew the mountains well, but the occupying forces didn't, and so the resistance were able to use the huge blocks of granite as cover.

Below: The gravel terraces left behind when the lake drained can still be seen above the houses here in Byrkjedal

Excursion 1.4:
Visit Gloppedalslura

To appreciate fully the size of this rockfall, it's best to see it for yourself. There is a car park and viewing area.

To reach Gloppedalslura:

By car: From Stavanger, take the E39 south. Just beyond Ålgård, turn left on Rv45. Follow Rv45 for about 31km, until you see the hotel at Byrkjedal on the right hand side of the road. Just past Byrkjedal, turn right along Fv 503. The car park is on the right hand side of the road at the top of the hill (made of scree!)

Access: car
Length: n/a
Climb: n/a
Grading: n/a
Facilities: none, but Byrkjedalstunet is close, with a cafe, toilets, hotel and gift shop.

Magma Geopark
Many more excursions!

The Geology trail at Eigerøy Fyr (Section 1.2) and rockfall at Gloppedalslura are both sites included in Magma Geopark. Magma Geopark is a large area to the south of Stavanger. It is a UNESCO Global Geopark, and includes some world-class geological locations, plus much more besides. The park is continually under development, with more sites and excursions being added.

For more information, visit their website at www.magmageopark.no. Here you will find recommendations for tours, and many excellent downloadable guides to help you find your way and explain the sites you are visiting.

Right: On the slopes of Vinjakula

Above: Evening over Lysefjord

Chapter 2:
It isn't all about Fjords

Tourists flock to Norway to see the fjord scenery: spectacular cliff edges, plunging waterfalls and ribbons of deep sea water winding along the bottom. Lysefjord, and Pulpit Rock (or Preikestolen as it is known in Norwegian) on its northern shore, are two of the most popular attractions near Stavanger. The fjords were created by glaciers, which over millennia scoured out deep trenches in the rocks that they flowed over. Now that the climate has warmed and the glaciers have retreated, the deep fjords have been left behind in all their glory.

Above: Pulpit Rock on Lysefjord

While fjords are the largest features created by glaciers, the ice has left behind many other traces. This chapter looks at some of these other legacies of ice and water, and where to find them.

Left: Lysefjord from Kjeragbolten

Above: Glacial striae near the summit of Undeknuten

2.1 Undeknuten
and the lines on the rock

The area of Sandnes municipality, or “kommune” in Norwegian, extends to the east and south far beyond the town itself, into the farm and moorland. The area contains a multitude of hills. All are loved and climbed, some more often than others. One of the less frequented tops is Undeknuten.

The route to the top of Undeknuten begins through farmland, but soon climbs to higher ground. Above the tree level, the terrain is boggy, with frequent patches of bare rock. The rock is smooth and, in some places, parallel grooves run across it. These are glacial striae: marks left by ancient glaciers. The land that we see in Rogaland today has been covered by ice fields and glaciers several times during the epochs.

Top: the summit of Undeknuten
Above: Lysefjord Bridge from the summit of Undeknuten

Looking down from the vantage point of an aeroplane, a glacier seems still, like a ribbon of ice winding along. Closer to the glacier, the reality becomes apparent. Where glaciers end abruptly at a fjord or in the sea, the summer sees regular "calving", a process through which chunks of the ice break away and float into the sea. The shoreline of the glacier remains in the same place though. The calving happens because ice from behind pushes the glacier slowly into the sea. The average glacier inches along at around a meter per day. Some are faster: the record is held by the Jacobshavn Isbrae in Greenland, which was clocked at 30 meters in one day. Others are more leisurely: at the top of an icefield, where the gradient is gentle, movement can be as little as half a meter a year.

Above: Glacial striae highlighted by water, Åmøy

If you have ever experienced the power of a river current, you are in a good position to appreciate the magnitude of the force slowly wielded by a river of ice. As glaciers flow, they pluck chunks from the rock that they pass over. These lumps and shards become embedded in the ice at the bottom of the glacier, and, as the glacier moves, they are both ground down into smooth boulders, rocks and pebbles, and, simultaneously, they scrape away at the rocks they are passing over. The incredibly deep fjords are created by eons of ice plucking away rock and grinding down the surface below.

Glacial striae are caused by erosion on a much smaller scale. They are the result of rocks being dragged along the surface of the bedrock, etching grooves as they go. Smaller particles act like sandpaper, and polish the rock surface. This leaves a "glacial pavement": smooth rock with grooves etched into it. Here in Rogaland, even the mountain tops have had ice flowing over them at one time or another. The picture overleaf, taken near the top of Undeknuten shows a stunning example of glacial striae, but they are visible in many, many more places around Stavanger.

Excursion 2.1:
Hike to Undeknuten

A straightforward hike through farm and moorland, with stunning views towards Lysefjord and Bynuten.

Access: car	
Length: 6.4km	
Climb: 315m	
Grading: moderate	
Facilities: none	

To reach the start of the hike:
From Stavanger, take the E39 south towards Kristiansand. Exit after about 15.5km on Rv13 towards Lauvik. Turn right after 3km, signposted Lauvik. After 15km, the car park is signposted on the left hand side of the road, just after the lake (Tengesdalsvatnet).

Hike Directions:
From the car park, cross over the road and then cross the stile that leads to the trail, signposted "Undeknuten". The trail begins along a farm track.
After about half a kilometre, as you reach a farm area, look for a signpost. "Rundtur" is signposted to the left, and "Undeknuten" straight ahead along the farm track. Take the left turn for "Rundtur", and follow the trail markings over the saddle and through the pasture on the other side. The trail is marked as it ascends a ridge and then joins a track near Rv 506. Climb up the track, and leave it at the signpost to continue up the ridge. The striae were spotted near the arrow on the map. Continue to the summit, and then follow the marked trail back to the car park.

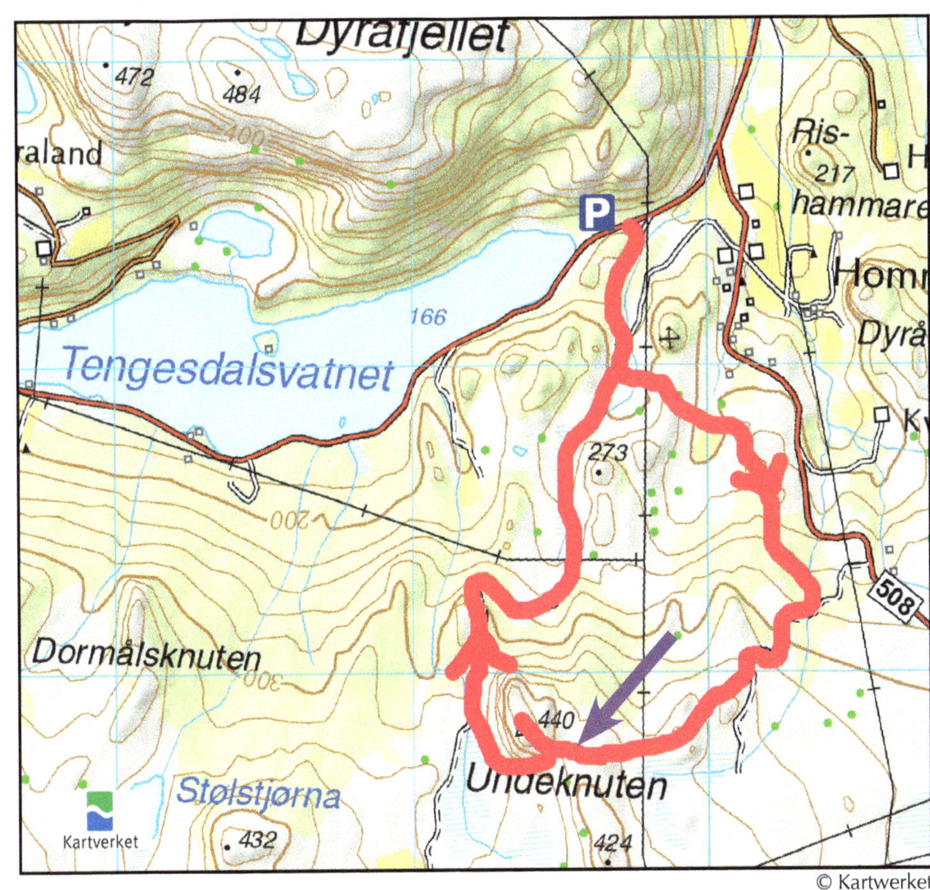

Above: Vassøy shoreliine

2.2 Vassøy: the exotic and the erratic

Vassøy is one of the "town islands", a group of islands close to Stavanger city centre. A series of bridges connects a chain of them to the mainland, but others that are smaller or slightly further away are reachable only by boat. Vassøy is one of the latter, served by passenger and car ferries, or reachable via private boat. A walking trail winds around the perimeter of the island, and on the north east lie some interesting rocks near the shoreline.

The rock that everything else (soil, pebbles, plants) is on top of is called bedrock. Here it is phyllite, a black shale with white veins of quartz running through it. Yet, close to the water, lie a series of large stones made of totally different rock. These are almost white, because of crystals of quartz and feldspar

Top: Vassøy has lovely flowers as well as rocks.
Above: The lighter-coloured exotic rock stands out from the darker bedrock on the shore of Vassøy

together with flecks of darker amphibole. They clearly didn't come from here. The nearest granite outcrops, places where granite is the underlying rock, are found many miles away, high in the mountains. Something has carried these rocks from there, and deposited them here. Rocks like this, that don't belong locally, are called "exotics", and these are a fine example.

Above: Flowers near the Vassøy shore

Further south along the coast, the trail rounds an enormous lump of rock. This time it's made of phyllite, the same as the bedrock, but it's still obviously been moved from elsewhere and dumped here. Around the island lie many of these boulders, some huge and others smaller: these are termed "erratics". They aren't just found on Vassøy. Erratics can be seen accross the whole of Rogaland, and form a normal part of the landscape in this part of the world. They are here because of the glaciers.

Glaciers are powerful enough to transport more than just small pieces of rock along with them. Sometimes, huge chunks are plucked from the mass of bedrock. Just like the smaller stones, they become embedded in the ice and are transported with it. When the ice melts, the boulders are left behind. Some of the boulders were left amongst the mass of debris deposited where the glacier ended, and others were left behind in the wake of retreating glaciers. This is how they are scattered so far and wide in the landscape today.

Left: an erratic on Vassøy

Excursion 2.2.1: **Hike around Vassøy**

An easy hike around the perimeter of the island, with fine views to Stavanger and Tau

Access: ferry
Length: 5.7km
Climb: negligible
Grading: easy
Facilities: picnic tables and BBQ pits at the beach

To reach the start of the hike:
From Stavanger, take a ferry from Fiskepiren (next door to the Magasin Blå shopping centre). Information about ferry times and fares can be found at www.kolumbus.no or at the information centre at Fiskepiren.

Hike Directions:
From the ferry dock, go along the road to the T-junction. Here, the trail is signposted to the left along a road. Follow the trail to the end of the road, and there look carefully for the trail markers that lead around a garden and on to the moor beyond the houses.
From here, the trail is marked with red dots on trees and rocks. Continue down to the shoreline and around the northern tip of the island. As the trail turns south, look for the exotics near the water. Follow the trail along the coast to find the huge erratic. Beyond the erratic the trail is well marked with T signs and red dots. A small beach with picnic tables, BBQ pits and a volleyball net offers a good place to picnic.

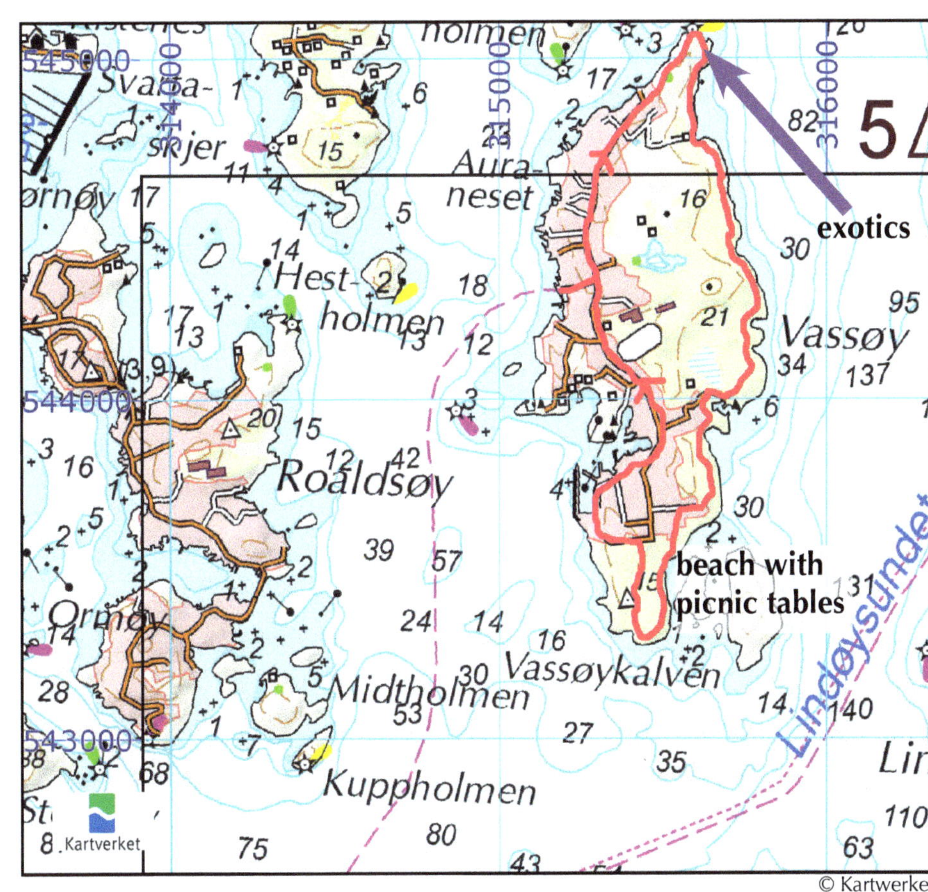

Excursion 2.2.2:
Hike to Jødestadfjellet and Resasteinen

This ridge walk is often neglected in favour of neighbouring Dalsnuten. It gives fun scrambling, airy views and fewer crowds. Resasteinen is a huge erratic with a book for visitors to sign.

Access: car or bus	
Length: about 6km	
Climb: about 350m	
Grading: moderate: steep and muddy in places	
Facilities: kiosk adjacent to the parking lot at Dale	

To reach the start of the hike:

By bus: Busses run from the centre of Sandnes to the start of the hike at Dale. For current routes and times, see www.kolumbus.no

By car: From Stavanger, take the E39 south towards Kristiansand and exit after 13 km on Rv509 towards Sandnes centre. Continue straight for about 1.5km. Just after going under a railway bridge, turn left, then second right and left again. Continue straight along this road until it becomes Daleveien. Drive along Daleveien with Gandsfjord on the left hand side. The road ends at the Dale buildings. Follow the signs to the parking lot.

Hike Directions:

Stand facing the information board in the car park and follow the sign to the right towards Resasteinen (the path straight ahead goes to Lifjell and the little beach at Dalsvågen). Follow the red markers left and over a stile to leave the Dale complex. The trail climbs the valley side and is rather muddy in places. Look carefully for trail markers, as the route forks left after about half a kilometre, and then turns sharp left up a slope and away from the track shortly before reaching the lake at Dalevatn.

At Dalevatn, the trail forks: to the

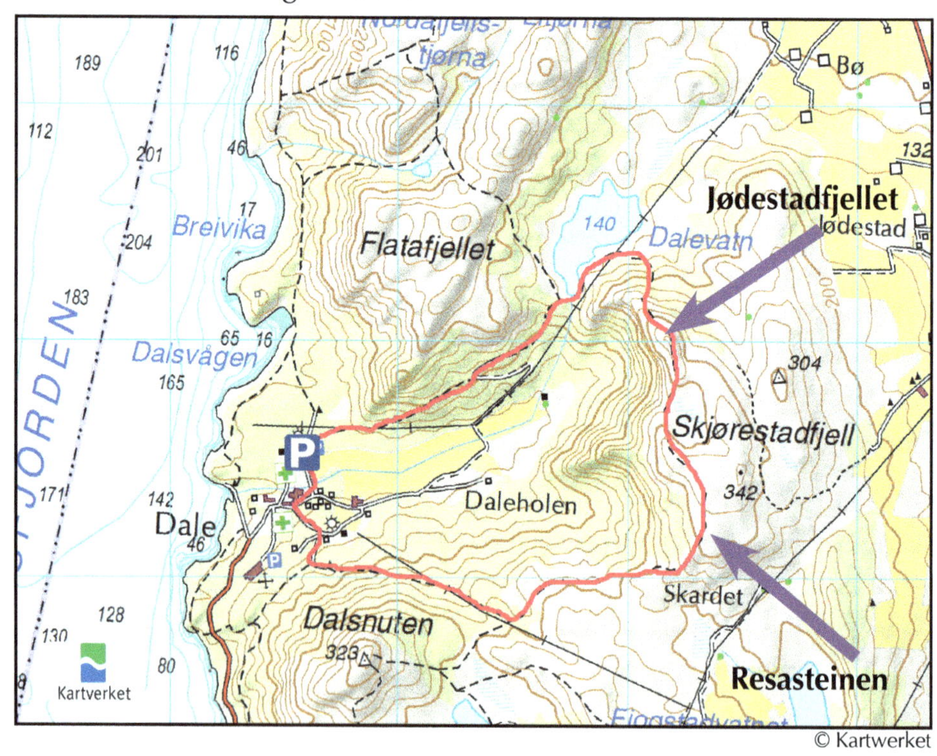

© Kartwerket

left is another route to Lifjell, whilst the trail to Resasteinen crosses the dam to the right. Cross the dam (it will be wet), and turn left along the side of the lake. Watch carefully for the trail markers as soon they turn up the hill and away from the water. This section is quite steep, but well marked, and after a little effort you will emerge on to the ridge of Jødestadfjellet.

Continue along the ridge, dropping down from the summit and then up again to a little top, before descending once again to Resasteinen. The visitor log book is located in the red mailbox under the lee of the boulder.

Beyond Resasteinen, the trail is marked as it goes down a rocky slope and into trees. Look for the signpost to the right indicating the trail returning to Dale.

Above: Resasteinen

More Erratics

Right: Steinfjellet

Above: Reinaknuten

Below: Hog-Jaeren

Left: The King's Stone, near Obrestad

Above: The shoreline near Reve

2.3 The Pebble Shorelines

To the south of Stavanger and Sandnes is an area called Jaeren. Jaeren isn't similar to the rest of Norway. Here, the land is reasonably flat, and instead of stones, mountain slopes and forests, there are fields and beaches.

The beaches are strung in a line down the coast, giving rise to the name "pearls on a string". They are beautiful indeed, with fine sand, long stretches of dunes, and waves from the North Sea. Many have camping sites, restaurants or other facilities, drawing the crowds both in fine and exhilarating weather.

Between the beaches is a different kind of coast. Sand gives way to pebbles of all sizes and shapes. Some of the beaches have pebbles well sorted into sizes, others show a

Top: pebbles at the beach
Above: Rounded rocks are often used for wall construction in Jaeren.

jumble of this and that: a large, rounded rock here, and a handful of fist-sized stones there.

The stones in these beaches were brought here by the outwash flows from glaciers further inland. During the Quaternary period, which is the past two million years, there have been several glaciation episodes. The Jaeren area was the boundary between a large flow that drained huge icefields in Scandinavia, and an ice sheet that covered the southwest of Norway. The beaches are the tip of the sediment: the Jaeren area is covered with layers of material from glaciers, interspersed with sediment from seas that existed inbetween the ages of glaciation.

This is why these beach pebbles are so beautifully rounded: they have been transported for hundreds of miles; all of that time, they were being tumbled and polished. Some of the rock eroded by the glaciers didn't make it to the beach as stones: much of it was ground to fine powder and gravel along the way. This too has been deposited in the Jaeren, and is called glacial till.

Above: Jaeren coast in August

The stones found here are plentiful, and have been put to good use. Field boundary walls and boathouses are built with them, but, even so, piles of stones in the fields are a testament to the amount of hard work that was necessary to clear the fields before they could be used for farming.

Left: Pebble shore south of Hellestø

Excursions for section 2.3:
Hikes between Hellestø and Orre

Beautiful coast between the beaches at Hellestø and Orre, easy to divide into day or half day hikes. Don't be daunted if transport is not available at both ends of the hike: things never look the same on the return journey, and there are always new things to see!

For more hikes along the Jaeren coast, see also Chapter 4.

Access	car
Distances	Hellestø to Sele: about 3.5km Sele to Reve Harbour: about 6.5km Reve Harbour to Orre Friluftshuset: about 4.5km
Climb	negligable, but some short, steep slopes on sand dunes
Grading	easy
Facilities	Toilets at Hellestø, toilet and Sunday cafe at Orre Friluftshuset

To reach the start of the hike:
Car parking is available at Hellestø beach, Sele, Reve Harbour (limited space) and Orre Friluftshuset.

Hellestø: From Stavanger, take the E39 south towards Kristiansand, and leave after 9.3km, signposted "Sola". After 5.3km, turn south (left) along Rv 510. Turn right along Nordsjøvegen after 2.5km, and then left after another 1.7km, signposted "Ølberg". At the T-junction after 2km turn left, and then right at the next T-junction. The car park is signposted on the right hand side of the road after another 2km.
Sele: From Stavanger, take the E39 south towards Kristiansand, and leave after 9.3km, signposted "Sola". After 5.3km, turn south (left) along Rv 510. After 9.2km, turn right along Selevegen. The car park

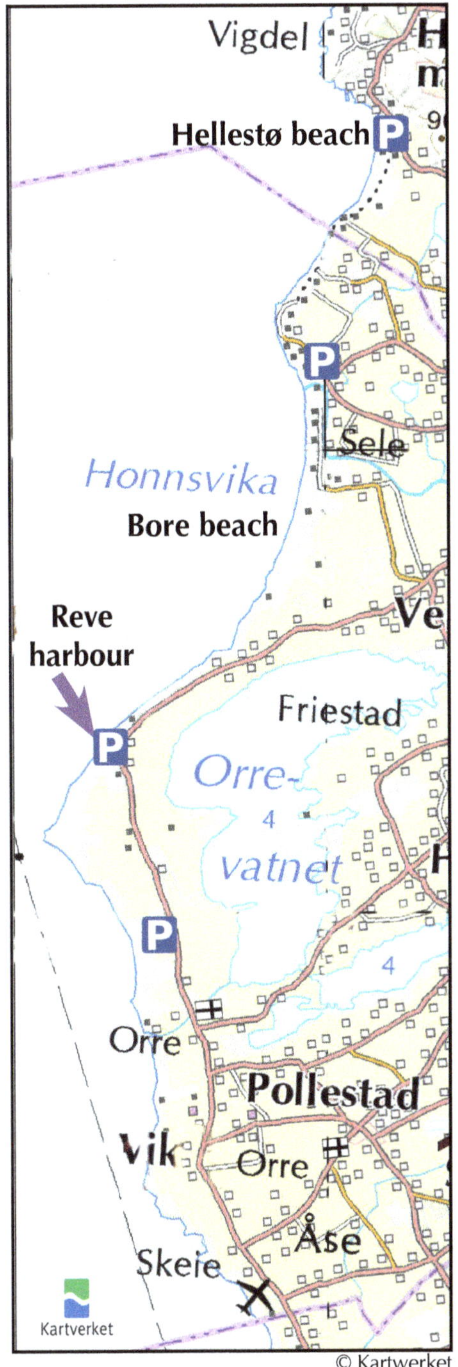

© Kartwerket

is where the road turns sharp left, near the shore.

Reve Harbour: From Stavanger, take the E39 south towards Kristiansand. After 12.2km, leave the motorway on Rv44 in the direction of Bryne. After 11.5km, turn north (right) along Rv510. follow Rv510 for 2.1km, and then turn left along Rv507. Reve Harbour is on the right hand side of the road after 5.4km, and there is limited parkig space here.

Orre beach: From Stavanger, take the E39 south towards Kristiansand. After 12.2km, leave the motorway on Rv44 in the direction of Bryne. After 11.5km, turn north (right) along Rv510. Follow Rv510 for 2.1km, and then turn left along Rv507. The car park is signposted on the right hand side of the road after about 9km.

Above: Summer flowers on the shoreline

Hike Notes:

Each of the beaches on this stretch of coast is a destination in its own right. Hellestø and Bore attract surfers and bodyboarders, Orre attracts families, especially on Sundays when the cafe at the Friluftshuset is open, and all of them are a mecca for kite flyers, dog walkers, and just about everyone. Much of this coast is also a nature reserve, attracting birdwatchers and naturalists. The coastal path runs along the beaches and the coastline inbetween them. The path is straightforward to find and navigate along.

This is a coast to explore in your own time and over several trips, parking at the car parks and doing "there and back" trips, or arranging for a car at each of two places and walking inbetween them. Enjoy the different experiences as the conditions change through the seasons and in varying weather. These are places you will want to return to again and again.

Right: Evening on the pebble shoreline

Above: At Reve, the shoreline is shaped like a hook, and provides a north-facing shelter from both south and westerly winds. There is a small harbour here, where boats can be tied in calm water away from the waves. An ancient burial mound nearby shows that this harbour has been used for milennia.

Above: Byfjord from the coast northwest of Mekjarvik. The E39 highway runs under this fjord in a tunnel, and the "whirlpool" feature opposite is just along the coast.

2.4 Kettles for Giants

Just north of Dusavik, the E39 highway disappears into a tunnel. Above the tunnel is a tranquil harbour. It's difficult to envisage the heavy traffic rushing through the tunnel beneath. From the harbour, the coast stretches northwest. Along the way is a beach, with a campsite adjacent, and beyond this a trail leads along the shore, towards the tip of the peninsula at Tungenes Fyr.

The rocks here are jagged phyllite. But, about a kilometre north along from the end of the campsite, just as the harbour at Tungenes comes into view, water swirls and splashes in a smooth, round crevice. It is quite deep, over a metre wide, and may have taken thousands of years to form. There are many more like it. These beautiful, smooth, rounded depressions may be found along rocky shorelines, in riverbeds, and where glaciers once flowed.

Above: The smooth crevice where water swirls along the coast from Dusavik, with a location map (top).

At the beginning of the twentieth century, there was some debate about what to call these rounded, smooth holes. "whirlpool" was suggested, but that refers to water, not rock. Another suggestion was "remoulade", which certainly sounds much more interesting than the rather banal alternative, "pothole". In English, it was the word "pothole" that stuck. At first this only referred to a natural feature in rock. Later, it also came to include that treacherous feature of modern tarmac roads: the hole that can appear almost overnight, especially in winter. That hole in the road is how we mostly think of a pothole in English today. The Norwegian language has a much more romantic name for a pothole in rock rather than in a road: here, it is a "Jettegryte", or literally "Giant's Kettle".

It is easy to imagine giant trolls cooking supper when you see the Jettegryte in Reianes on the island of Rennesøy. Here, the pothole was formed when sea levels were higher, and now the sea has retreated, leaving the pothole some way up the coast for all to admire. It is signposted just off the southern part of the Reianes peninsula trail (see Excursion 2.4.1).
For a pothole, the Reianes example is quite little, and there are larger "giants' kettles" to find. Near Mjuhaug in Kvernevik, there are

Below: The pothole at Reianes: light reflecting on the water inside shows how deep it is.

Fig 2.4: Formation of a Pothole

① 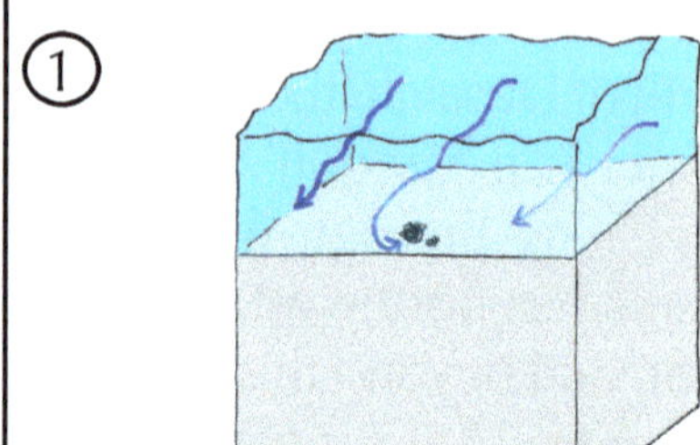

Vertical eddies, or mini-currents form in flowing water, often around an obstacle on the bedrock.

② 

The currents stir rocks around in a circle. The rocks are called "grinders", because they grind away at the bedrock below, forming a hollow.

③ 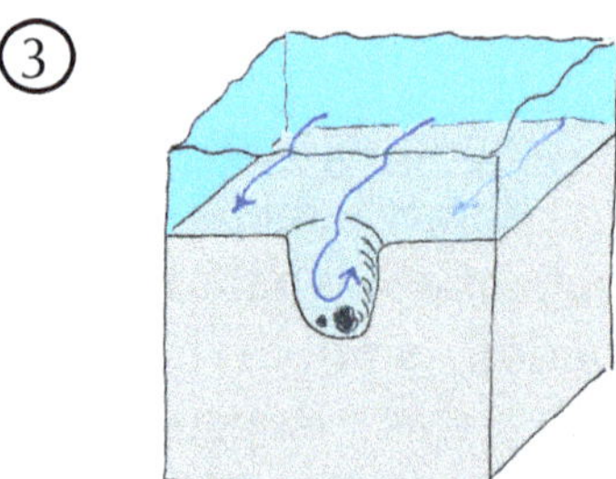

The vertical currents persist, and continue to swirl the grinders around the hollow, eroding the hole deeper and deeper over time.

two. The first is around 100m west from the top of Mjuhaug. It is an almost perfect circle, and modern "giants", or even normal sized barbecue chefs use it as a large, natural firepit. Further towards the coast, the sea has formed a horizontal pothole along a weak line in the rock. This cave is called Skrubbhåla. In 1940, people hid here, afraid of the Nazi occupiers.

Near the foot of Lysefjord lies a little island. To one side of the island is a truly huge pothole. According to legend, this is where local residents would bathe on a Sunday, before going to church. Other large potholes have other stories. On the road to Kristiansand, a huge "giant's kettle" is said to have formed by the side of the road where St. Olaf turned his horse around. The legends are impressive, but the same geographers who debated the naming of potholes wondered how such large ones could possibly have formed.

As with so many of the enormous land features in this area, we're back to glaciers. As the glacial ice slowly moves, giant cracks, or crevasses, form. During warmer days, some surface ice melts, forming rivulets at the top and sides of the glacier. When a rivulet meets a crevasse, the water plunges down, sometimes for hundreds of feet. At the base of the glacier is a layer of rocks and pebbles, plucked from the bedrock and being transported and worn down as they move. When the water from the rivulet hits these stones at high velocity, the stones are stirred around. Volumes of water, hitting these stones with huge energy, generates vortices, stirring up a circular storm of gravel and rocks. The gravel and rocks scrape away at the bedrock, digging the huge pothole.

Below: The pothole at Mjuhaug

© Kartwerket

Above: Skrubbhåla: note the different types of rock on either side: this is due to the fault line here.

Coastal and river potholes form in the same way: swirling water stirs pebbles, which grind away at the rock below. Over time, a shallow depression becomes a hole, and another "giant's kettle" is born. The largest potholes in Northern Europe are also found in Norway, on the south coast at Slid. By the banks of the river at Nissedal, in the east of Norway, is another set of huge potholes. In both places, the water is relatively shallow and heated by the rocks, forming popular summer swimming areas.

Above: Time and falling sea levels have left the Reianes pothole well above today's shoreline.

Below: The pothole at Lysefjord

Excursion 2.4.1:
Hike around Reianes

An easy hike around this peninsula, with fine views and a pothole. In spring, this is a fine place to find primroses.

Access: car	
Length: 4.6km	
Climb: negligible	
Grading: easy	
Facilities: none	

To reach the start of the hike:
From Stavanger, take the E39 north towrds Bergen. About 4.9km after the second tunnel (Mastrafjordtunnelen), turn left. If you reach the ferry terminal, you have overshot. The car park is on the right hand side of the road after about 2.4km.

Hike Directions:
The hike is well marked with red dots and T-markers. From the car park, turn right and proceed a short way up the hill until the route is signposted off the road from the right hand side.
From here, the trail skirts the peninsula, although after the first kiometre is is a little distance away from the shoreline.
The pothole is signposted "Jettergryte" from the main trail, and is about 25m to the south (marked by the purple arrow on the map). Look carefully for the sign above the hole, as it is not easy to see from the side!
Beyond the pothole, the trail continues back to the road at Låde, with an optional detour to an ancient burial mound and viewpoint. From Låde, follow the road back to the car park.

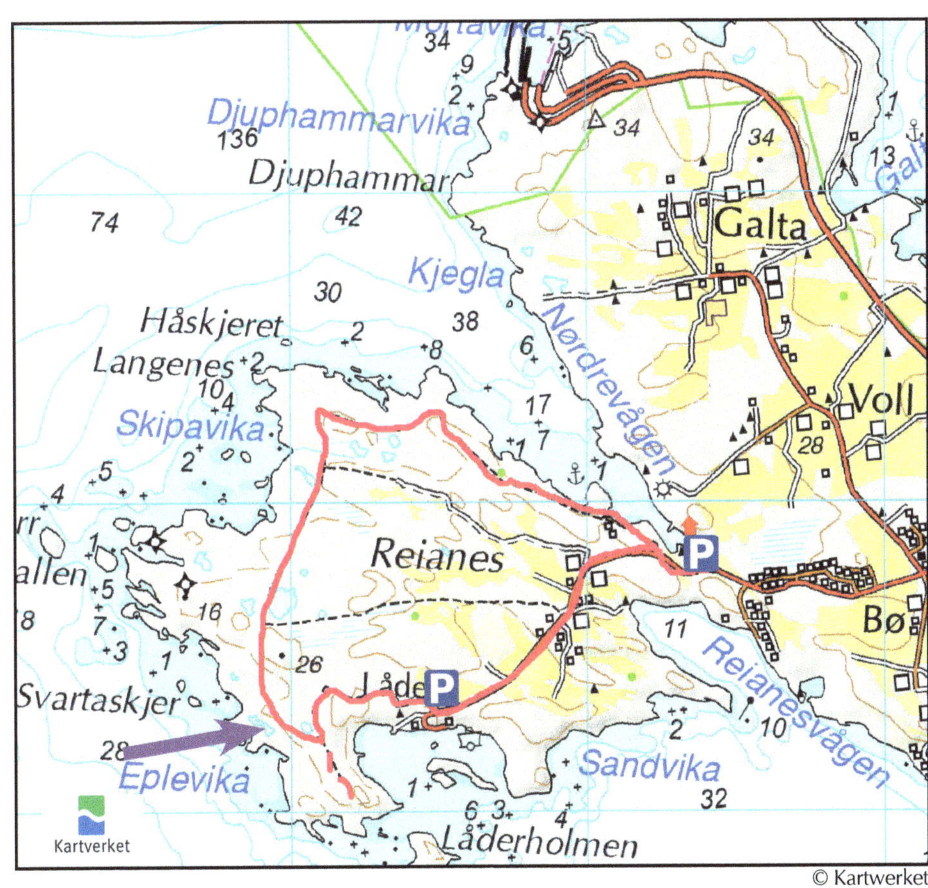

© Kartwerket

Excursion 2.4.2:
Bicycle ride from Hafsfjord Bridge to Tungenes

A fun bicycle ride along gravel tracks and quiet roads, with many places to visit along the way.

Access: bicycle or bus

Length: approximately 15km (in one direction)

Climb: negligible

Grading: easy

Facilities: toilets and Sunday cafe at Tungenes Fyr

To reach the start of the ride:

From Stavanger centre, take Rv 509 (Madlaveien) west for about 8km. The ride begins just before the bridge (Hafrsfjord Bridge). This starting point is easy to reach from many places in Stavanger, but from further away, it may be preferable to journey to the start of the ride by bus. Bicycles are allowed on buses, provided there is space. For current routs and times, see www.kolumbus.no.

Ride Directions:

The ride is marked reasonably well with signposts. It begins at the shoreline just underneath the bridge. From here, stand facing the fjord and begin by turning right, so that you are riding with the water on your left.

Ride along a short section (about 600 metres) of the Kvernevik Ring Road, and then turn left along a gravel track. The pothole at Skrubbhåla can be seen from the first little harbour on the track, about 300 metres from the road. The pothole can be seen at the shoreline by turning and looking back. Its approximate location is marked by the purple arrow on the map.

The ride continues around the Kvernevik coastline and through a short stretch of houses to Viste beach and the Viste coast. Just after a small harbour advertising boats for rent, look for the signpost indicating a right turn uphill and away from the shore. The ride now follows local roads for about 3.3km, until a signposted left turn along a farm track. The track cuts accross to another road section along Bøveien and then left along Sandeveien. At the T-junction, turn left to follow

© Kartwerket

Tungenesveien to Tungenes harbour.

Above: Tungenes Fyr

To visit the lighthouse at Tungenes Fyr, follow the sign to the left. Tungenes Fyr has a Sunday cafe, concerts on some evenings, and often boasts art exibitions in the grounds and inside the lighthouse.

To visit the "whirlpool" featured at the start of this section, leave the bicycle at the harbour, and follow the marked trail up a little slope to the south east (a right turn if you are facing the harbour from the road). The trail follows the rocky shoreline, and the feature is about 300m along.

For an alternative route back to Stavanger, follow the signposted bike route to Randaberg and on towards Tasta. Just beyond the junction at Gabbas butcher's, cross the E39 at the pedestrian crossing and take the signposted track to Stokka lake. Ride around the lake shore in either direction as far as the south side, and take the track to the car park near Madla Amfi. From here, either take the bus or cycle along the Madlaveien cycle route back into town.

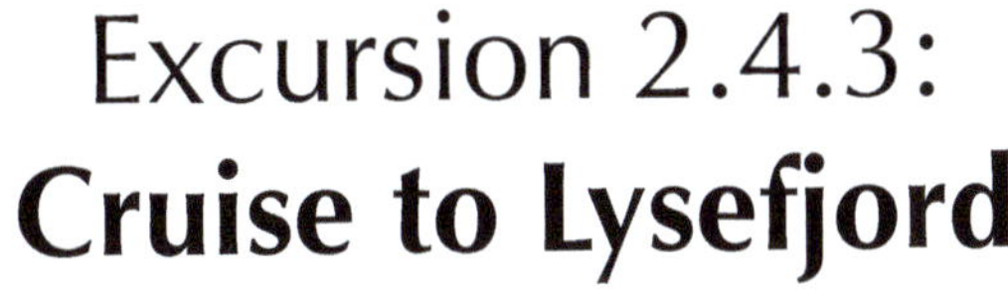

Excursion 2.4.3: **Cruise to Lysefjord**

See the sights of Lysefjord, including the pothole, the relaxing way, from the deck of a boat.

Above: Lysefjord

During the summer months, and on Saturdays all year, half day sighseeing trips to Lysefjord/Pulpit Rock depart from Vågen in the cenre of Stavanger. The boats are modern and comfortable, and the tour visits several places along Lysefjord, including (usually) the pothole.

For current schedules and prices, visit www.rodne.no. During busy times, the trips can be popular, so advance booking is recommended.

Above: Standing Stone, near Madlamark

Chapter 3:
Of Monuments and Mysteries

Madla is a thriving suburb. Located close to Stavanger city centre, but still within easy reach of Forus, Tananger and Sola, it's easy to see why living here is popular. Madla has been built up in recent decades, but there are still some green areas. One of these is a hill by the side of three large blocks of flats.

From the top of the hill there is a commanding view of Hafrsfjord, and there are paths there from several directions. If you take the one from the north, you will find yourself going past a standing stone (pictured left). The stone has remained standing while centuries of storms battered it and decades of evening strollers have ignored it. Older than written history, it is not the only ancient monument in the area.

Scattered all around the coast are traces left behind by people who knew these fjords and hills long before us. Midden (rubbish) heaps show us their waste, and allow us to figure out what they ate. Tools show us how they caught (or farmed) their food, and burial mounds show us that respect for the dead goes back a long way. We're constantly finding, and learning, more.

Above: Rock carvings at Solbakk, near Tau

Left: Burial mound, Reianes

Above: Solbakk, near Tau

3.1 The Petroglyphs: Rock Carving by the shore in the Bronze Age

In 1879, a strange discovery was made near Revheim. Here, close to Hafrsfjord, there is a rock face that has been smoothed by the passing of one or more glaciers. On the rocks, there were some indented figures. There were boats, hand and foot shapes, circles and dots: certainly not anything that could have appeared naturally. Someone had taken the time to chisel the figures very carefully into the rock.

The rock carvings, or petroglyphs, date back to the Bronze Age, about 1800-500 BC. They were carved gradually, one or a few at a time, over a span of hundreds of years. They cover several rock faces.

Today the site of the carvings is a quiet place. The busy road to Tananger is some distance away. During the Bronze age, there was

Top: Rock carving at Solbakk.
Above: Dusk at the shoreline in Åmøy. The rock carvings are located above this shoreline

bog or forest inland, and the busy road of the time was Hafrsfjord itself. Boats were the main means of long distance transport and, sure enough, many of the carvings depict boats.

The petroglyphs are the result of hours of painstaking labour, so this site at Revheim that is so quiet today must have been very important in the Bronze age. In 1894, two bronze helmets (now on display in Stavanger Archaeological Museum) were found 200 metres southeast of the petroglyph site. Such a find is incredibly rare, and it suggests that Fluberget was a very important religious or ceremonial site.
There are a few other petroglyph sites around the Stavanger area. All of them have views towards the water. Two of the larger sites are at Austre Åmøy and at Solbakk, near Tau.

The name "Solbakk" gives part of the game away. "Sol" means "sun" in Norwegian, and "bakk" means "hill", so the name translates to something like "sunny hill". The area is sheltered and facing in the right direction to catch the sun. There are several traces in the

Above: A boat symbol at Fluberget. Boats were the main means of long-distance transport, and it is believed that they were very important in religious beliefs and legends: the sky was seen as an ocean, over which the most important gods sailed in their boats.

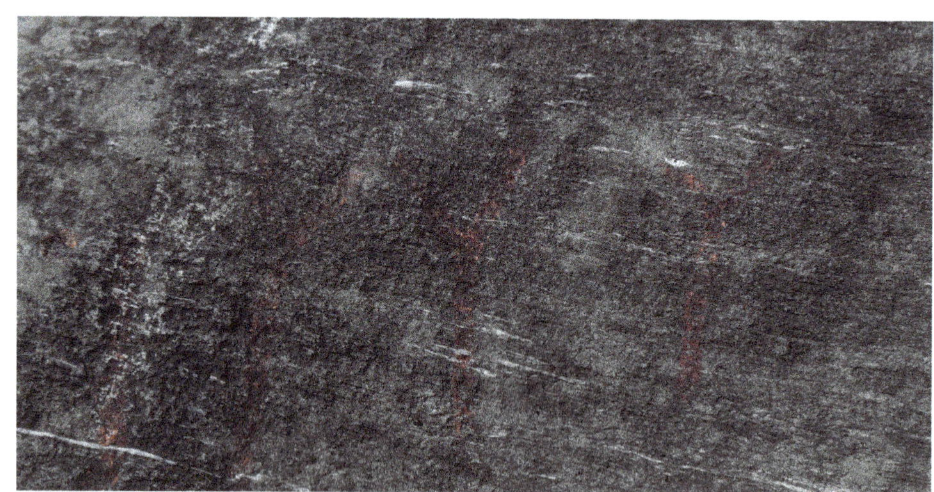

Above: Perhaps these Y-shaped symbols were intended to represent people, with their hands raised to the heavens.

Above: These foot symbols could be the feet of the gods, walking on the earth. The hand symbols might also represent those of the gods.

Above: Circles, rings and spirals are taken as being the sun.

neighbourhood of the ancient settlement here. The petroglyphs are carved into a rock face by the shore.

Again, we see boats and sun symbols, but here there are two distinct kinds of boats. One kind of boat is shorter, has a simple hull and different shape, whilst the other has a double hull and is clearly more sophisticated. Experts believe that these symbols were carved at two distinct times, with the different boat carvings corresponding to the boat technology of the day. The simple hulled boats may have represented skin boats, and if this is correct, they were carved at a much earlier date than the double hulled boats.

Above: The meaning of the ladders and frames (bottom left) is lost to time.

The double hulled boats are the only petroglyphs of boats that have been found so far at Austre Åmøy. Austre Åmøy is the largest of the areas of carvings around Stavanger. Although not all of the carvings are accessible to visitors (this is a working farm), in all there are about 1000 carvings, spread over an area of about 1km, in 10 separate areas. They include the largest boat carving in Norway, at about 5.5 metres long.

The petroglyphs were difficult to see except in very low angled light, so they have recently been picked out with red paint, so that we may all see them. We may never know exactly why they were carved or exactly what all the symbols mean, but we can imagine these quiet places by the shore bustling with people on ceremonial days, as boats arrived and departed.

Excursion 3.1.1:
The Revheim Tour

Visit Fluberget, the Hafrsfjord shore and Stavanger Golf course on this pleasant circular walk.

Access: car or bus
Length: 7.8km, or about 8.8km from the car park at Stokkavatnet
Climb: negligible
Grading: easy, suitable for pushchairs and wheelchairs
Facilities: cafe at the golf course

To get to the start of the walk:

By car: Although there is no car park on this walk, there is one close by just south of Store Stokkavatnet. From Stavanger, take Rv 509 (Madlaveien) towards Tananger. The car park is signposted on the right hand side of the road after about 3.5km.

By bus: Buses run frequently along Rv 509. The bus stops by the church at Revheim and at Sundetunet are on this walk. For current schedules and times, see www.kolumbus.no

Walk Directions:

This walk is taken from the "Stavanger 52 Walks" series. It is very well signposted, with T-markers and "no 23 Revheimsturen". A large scale map is available for printing from the Stavanger Kommune website: www.stavanger.kommune.no

To reach the hike from the car

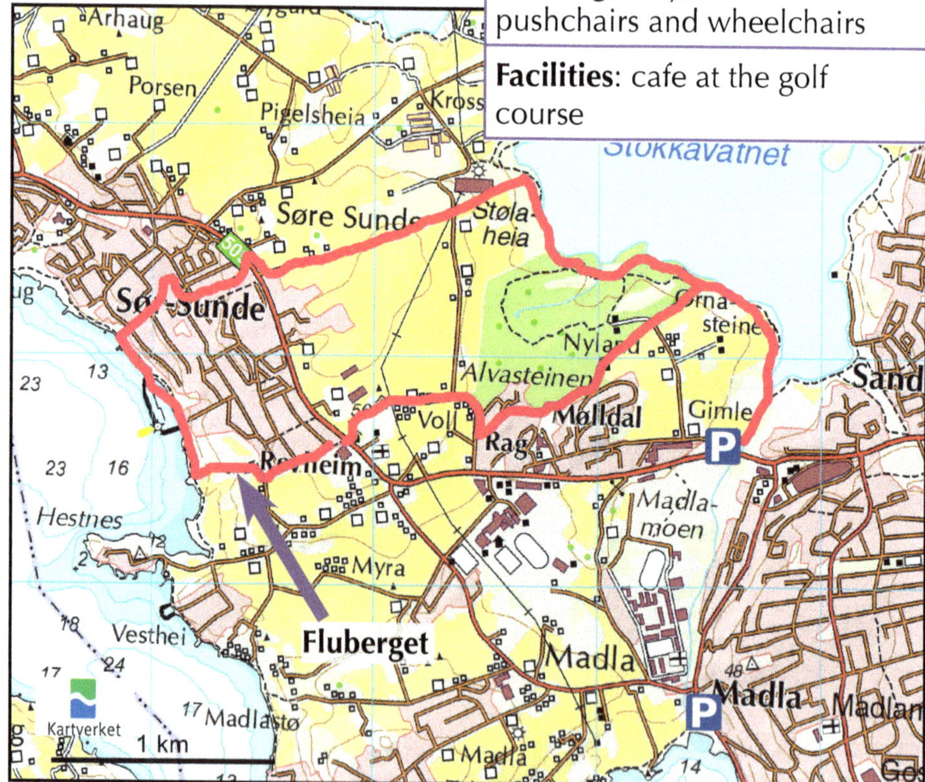

© Kartverket

park to the south of Store Stokkavatnet: Stand and face the red mural on the wall of the electricity substation by the side of the car park. Look to the right hand side of the electricity substation, and you will see a track leading past the building and away from the car park. Take this track. Afer about 200m, turn left at the T-junction.

Just as you reach the golf course, look for the T-marker directing "Revheimsturen" to the left. From here follow the markers around the walk.

Excursion 3.1.2:

Visit the Petroglyphs at Solbakk

This sunny site, with its view towards Stavanger, makes a great stop when there is time available while waiting for a ferry.

To get to the site at Solbakk:

From Jørpeland: Take Rv13 north towards Tau. The site is signposted to the left of the road after just under 7km. There is a small parking area by by the side of the road.

From Tau: Take Rv13 south towards Jørpeland. The site is signposted to the right hand side of the road after about 4.7km. There is a small parking area by the side of the road.

From the car park, the short path to the petrogylphs is signposted.

Below: Suns and ships at Solbakk

Excursion 3.1.3:
Visit the Petroglyphs at Rudlå

At the time of writing, the paint had almost completely faded from this little site, leaving the visitor with the same feeling of looking for the carvings that their original discovered must have had.

Above: The outline of a boat at Rudlå

To get to the site at Rudlå:
Rudlå is just 10 minutes on foot from the centre of Stavanger. From the centre, stand facing the main cathedral entrance, turn right, and walk along the road until you reach a T junction. Turn right, and continue up the hill towards Lokkeveien. At Lokkeveien turn right, and continue to the junction with Stokkaveien. Turn left up Stokkaveien for about 200m, and just past the playground on the right hand side, turn right along the path past the entrance to a white building. The site is on the left just past the building, between it and a nursery.

Above: Glacial striae at Rudlå (see Chapter 2).

Excursion 3.1.4:
Visit the Petroglyphs at Austre Åmøy

The largest area of petroglyphs in Rogaland is just a short journey from Stavanger by car.

To get to the site at Austre Åmøy:
By car: From Stavanger, take the E39 north. After exiting the first tunnel (Byfjordtunnelen), look for a turning to the right. Take this turning (Åmøyveien). If you cross a bridge, you have gone too far. follow Åmøyveien for about 7km, to where it ends at a little harbour. There is limited space for parking here.
To reach the petroglyph fields, walk back up the road towards the farm buildings. There you will find a noticeboard with information and a map showing where the sites that are open to the public are. Please remember that these are working farms and do not disturb animals or crops.

Below: Boat carving at Austre Åmøy

3.2 Standing Stones: and what becomes of them

The standing stone at Ullandhaug at the beginning of this chapter marks a grave from the Bronze Age. Stones have been placed alongside it in the form of a cross. There are many other standing stones along the coastal areas where prehistoric people lived in Rogaland. The stones have clearly been quarried and shaped carefully. The sides of most have been squared, and some taper beautifully towards the top of the stone. Once quarried and shaped, the stones must have been lugged with great difficulty to their final standing places. Given that many are still standing today, they must also have been placed well into the ground, with support where necessary, meaning that a good deal of digging was involved.

Top: Was this gatepost made from a standing stone?
Above: Three standing stones near the shore of Hafrsfjord at Sunde

Sometimes there is one standing stone, sometimes three, occasionally even more. Burial remains tell us that some of these stones were grave markers, but there are other standing stones that are not marking graves, and the meaning of many of these is lost to time. Time also has changed many of the sites where the standing stones are placed. They must originally have been at important sites or graves. Many of the sites where they are found have commanding views: they were carefully chosen.

But, as time has moved on, the meaning of the stones and their sites has fallen out of cultural memory. Brush and undergrowth have covered over some of the sites so that now they are difficult to find. Other stones have been overtaken by the city, and now are part of gardens or parks. One site seems literally to have been buried in sand, like an ancient Egyptian monument.

Slightly to the south of Sola airport there is a circle of standing stones. There are other stone circles elsewhere in Europe, for instance those at Stonehenge in England and Callanish on the island of Lewis in Scotland, but stone circles are unusual in Norway. It is near the beaches at Sola and Rege, in an area that oral history describes as having been affected by shifting sands in the past. This stone circle was described in 1745 by Bendix Christian de Fine. He wrote of a circle of 24 upright gray stones, with a stone in the middle that was horizontal, like a table, and smaller, white stones in-between them. There aren't records of it before this, so perhaps the circle had, up until this time, been fully or partially buried, but the sand had eroded away quickly to expose it.

The original significance of the site remains a mystery, its meaning lost generations ago. With no reason to preserve the circle, the locals found the stones to be very useful indeed. By 1879, Anders Lorange, a conservator from the Bergen Museum, only found two of the large standing stones remaining in the circle. The others had been taken to new places. One had been put to use in a drying-house at Ølberg, whilst another was part of a wall in a boathouse at nearby

Above: This standing stone is now in the car park at Soma Gård, near Sandnes, but the series of holes drilled into one side of it betrays the fact that is has been put to another use at some time in its history.

Above: The reconstructed stone circle near Sola Aprport

Raegestranden. The owners of the boathouse gave permission for the stone to be returned to Sola, but the owners of the drying-house were not prepared to give up the stone they were using. Other stones had been put to use in many ways, such as fence posts, gateposts or road markers, in the same way that good building stones have been "upcycled" all over the world for milennia.

It was another century before anyone became interested in reconstructing the stone circle. Eventually, a local historian, Kari Myklebust, began to clear away the overgrowth at the site of the stones. Others became involved, and with the help of local farmers, many of the standing stones were recovered. In 2008, reconstruction of the stone circle was complete. To this day, nobody really knows why the stone circle was built or for what it was used, but those wishing to speculate or just enjoy the site may visit.

Nearby, along the west side of Hafrsfjord, there is a series of standing stones, said to mark places in the saga of Erling Skjalgsson. Erling Skjalgsson was a Jaeren lord, ruling over an area in the southwest of Norway. Following several quarrels with King Olav, Erling was slain in Soknasundet on 21 December 1028. Snorre told many tales of Erling Skjalgsson in his sagas, and he links each of them to a standing stone in the line along Hafrsfjord. The stone at the north of the line is beside Rv510 at the south side of the Hafrsfjord Bridge. Anybody crossing the fjord, even before the bridge was built, would have seen the standing stone here. It is marking a pre-Christian grave, perhaps from Viking times.

Above: The grave marker at Jåsund, attributed to Erling Skjalgsson.

The advent of runic writing reveals the mystery of some of the later standing stones, since the people who erected them were now able to carve upon the stone their names and the reason why the stone was placed. The island of Rennesøy is home to several standing stones, and three rune stones have been found there in the mediaeval church at Sørbø. One of these was hidden above a window in the church, and reads:

"Tormod and Torgard erected this stone in memory of Gaut"

Another runestone was discovered in the foundations of St. Mary's Church, and is now in the Stavanger Archaeology Museum. The church has been demolished, although the outline of its foundations is preserved by the side of Stavanger Cathedral. This stones is thought to date to around 1000 AD, and the runic inscription upon it has been translated as:

"Ketil erected this stone in memory of Jorunn his wife, Utyrmes daughter."

Above: The view looking west from Tinghaug towards the North Sea.

Some memorial stones were not placed at sites of worship. A runestone was found in the River Sokna, which had originally been placed by a bridge. It has upon it a very early monotheist inscription, dated to the first half of the eleventh century:

"Sakse made, giving thanks to God, this bridge for the soul of his mother, Turid".

A runestone placed at Sele, dated to 1100 AD, on the Jaeren coast, shows us where the phrase "carved in stone" comes from:

"The agreement upon this stone is that half the fishing area belongs to (Sele) as a hereditary right".

Inland from Sele, there is a low ridge that rises from the Jaeren plain. The highest point in Kleppe Kommune, at 104 meters above sea level, is called Tinghaug. In prehistoric times, this area had everything: fertile soil, a commanding view for miles around in all directions, and quick links to the sea. It hosted a thriving community, and excavations have shown that there are several important sites around the hill. A large grave at Grønhaug is marked by a standing stone. This standing stone is not ancient: in 1879 Lorange (see earlier) excavated the site. When he was finished, he continued the tradition of marking gravestones with standing stones by erecting a standing stone there.

Further down the hillside is another grave site, called Krossberget. The grave was originally occupied by a lady, and, judging by the goods she was buried with, she was very high status indeed. About 600 years after the grave was built, a standing stone was erected on the site, but now the standing stone is a new shape: it is a cross.

Excursion 3.2.1:
Visit Stavanger Archaeological Museum

This chapter covers only a tiny fraction of the information available about prehistoric times in the Stavanger area. The best place to begin to find out more is the Archaeological Museum. Here, you will literally find a treasure trove, including the runestones mentioned in this section. More information, together with location and opening hours, is on their website at: am.uis.no

Above: The Iron Age Farm at Ullandhaug

Excursion 3.2.2:
Visit the Iron Age Farm at Ullandhaug

The reconstructed farm is open during the summer season.

To get to the Iron Age Farm:
By bus: the nearest bus stops are at Gosen and the University of Stavanger. For current routes and times, see www.kolumbus.no.
By car: the address of the Iron Age Farm is Ullandhaugveien 165, Stavanger. Parking is extremely limited, but there is a little car park quite close at the Botanical Gardens.

Excursion 3.2.3:
Visit the Stone Circle near Sola Airport

The stone circle has been beautifully reconstructed.

To get to the Stone Circle:
By car: From the centre of Stavanger, take the E39 south towards Kristiandsand. After about 6.5km, exit towards Sola airport. After about 4.4km, turn left towards the airport, and then after about 1.2km, turn right, signposted "Sola Strand Hotel". The road to the car park is signposted to the left after about 2.7km. The car park is on the left hand side of the access track. The stone circle is beside the car park.

Below: Detail of the reconstructed stone circle near Sola.

Excursion 3.2.4:
Visit the site at Tinghaug and Krossberget

Access: car
Length: n/a
Climb: negligible
Grading: easy, but not suitable for pushchairs and wheelchairs
Facilities: none

The view alone makes this site well worth a visit. It would make a good cycle expedition from Stavanger, Sandnes or one of the stations along the Jaeren train line. It is also on a circular walk signposted by Kleppe kommune. The walk is about 8km long: for details see www.kleppe.kommune.no "Tinghaugrunden Tur".

To get to the site:

By car: From Stavanger, take the E39 south towards Kristiansand. After about 12km, exit on to Rv44 towards Bryne. Continue along Rv44. You will bypass the town of Kleppe, and go through a short tunnel. About 16km after leaving the E39, and just opposite a bakery, and just before entering the town of Bryne, turn left (signposted Tu). If you find yourself in the town of Bryne, you have gone too far.

After turning left off Rv44, follow the road past Tu Barnehage. Shortly after this, take a left turn up a track, signposted to Tingberget. Go up the short hill, and park in the signposted car park on the left hand side of the track. There are several information boards in Norwegian around the site, plus one in English at the car park.

Above: The Cross at Krossberget

Left: The larger of the two crosses at Tjora

3.3 The Standing Crosses

It is thought that the stone cross above the grave at Krossberg was placed there about 600 years after the grave was dug, in around 1000 AD. This makes it one of the oldest Christian monuments in the area. The cross was ruined during the St. Hans celebrations of 1855, but the pieces of the cross were put back together and raised again in 1937. There are other very early crosses still standing at Kvitsøy, Tjora, and one now in the Stavanger Museum, which has an interesting story.

In addition to the standing stones along the west side of Hafrsfjord that are associated with the story of Efling Skjalgsson (see Section 3.2), a cross was raised with a runic inscription attributed to him. This cross was originally erected in the Viking settlement at Stavanger, where the road from Jaeren entered the town next to the present day Lake Breiavannet.

Top: Front view of a cross at Tjora
Above: Both crosses side by side at Tjora

The runic inscription on the cross tells that it was raised by Alfgeir, the Priest, and that Erling was his lord. Since it was first raised, the cross has been moved three times, the last of which was to Stavanger Museum, where it is now a treasured monument. A reconstruction of the cross has been erected at Sømme, the last in line of the "Erling-run" of standing stones. Just a short distance away from Sømme is the ancient site at Tjora.

Tjora is a very early Christian site, with two crosses, which stand today in the graveyard where there was also once a wooden, or stave church. The church here is recorded from 1270 AD, but there is evidence that Tjora was used for worship earlier than this. A 2.3 metre long gravestone was found in the wall of the churchyard. It has upon it a runic inscription, and has been dated to about 1150 AD, so the site must have been in use by then.

Originally, there were 4 stone crosses at or near Tjora dating to between 1000 and 1100 AD. After the church was demolished, the crosses remained intact until some time between 1870 and 1880, which is when two of them were sent to Bergen. One of the remaining crosses was destroyed in an accident, and the last one destroyed during the Second World War. These two crosses have now been reconstructed, and stand once again at Tjora. One of them has a similar form to the cross in Stavanger Museum associated with Erling Skjalgsson, so it is assumed that it comes from a similar date, very early in the days of Christianity in Norway.

It may be inferred from place names containing the word "kross" that other crosses once stood at holy sites around Rogaland. One of these places is Krossberg, which is between Store Stokkavatnet and Hålandsvatnet. Here, oral tradition had told of a stone cross on the knoll still known as Krossberg. In the 1920s, someone looked for a ruined cross there, but couldn't find it. They scoured the area, and eventually fragments of the top of a stone cross were found at the side of Hålandsvatnet a hundred metres from the knoll. The fragments are now safely housed in the Archaeological Museum. In 2000, local

Above: The modern standing cross at Krossberget

residents commissioned a granite copy of the cross, and placed it where the original cross is thought to have stood a thousand years earlier. Beside the new cross is a brass plaque, with engravings of all the known early stone crosses in the area upon it.

Even though the top of the original cross was discovered, there is still a mystery. At no time in the past milennium has the significance of a standing cross been forgotten in local culture. It's difficult to see how anyone would have reused the stone from a standing cross for anything else, except possibly if they were desperate. How, then, was the cross broken? What happened to the stem of it, which still has not been found? Who carried the top of the cross to Hålandsvatnet, and why?

Below: The site at Krossberget, showing the bronze plaque with the depictions of the local standing crosses upon it.

Excursion 3.3.1:
Cycle around Hafrsfjord

Hafrsfjord, being sheltered with a long length of shore, has provided a haven to hunter gatherers, farmers, and Vikings. Visit sites left by these successive cultures along this pleasant cycle route, and feel how important this fjord must have been.

Access: car, bus or cycle
Length: about 23km
Climb: mostly flat
Grading: easy
Facilities: kiosk and toilet at Møllebukta

Here, the description is written from the car park at Møllebukta (the Three Swords beach), in an anti-clockwise direction, but it is possible to start the tour at any point, and cycle in either direction.

To get to the start of the ride:
From Stavanger, take Rv 509 (Madlaveien) towards Tananger. After about 3.6km, turn left towards Sola. The car park is signposted on the right hand side of the road after about 1km.

Ride Directions:
From the car park, stand facing the water and look to the right of the beach for a track leading along the shore. Take this track (the fjord should be on your left hand side).
The track is signposted along shoreline, through the new path cutting (Section 1.1), and over the new "bridge'. After the "bridge",

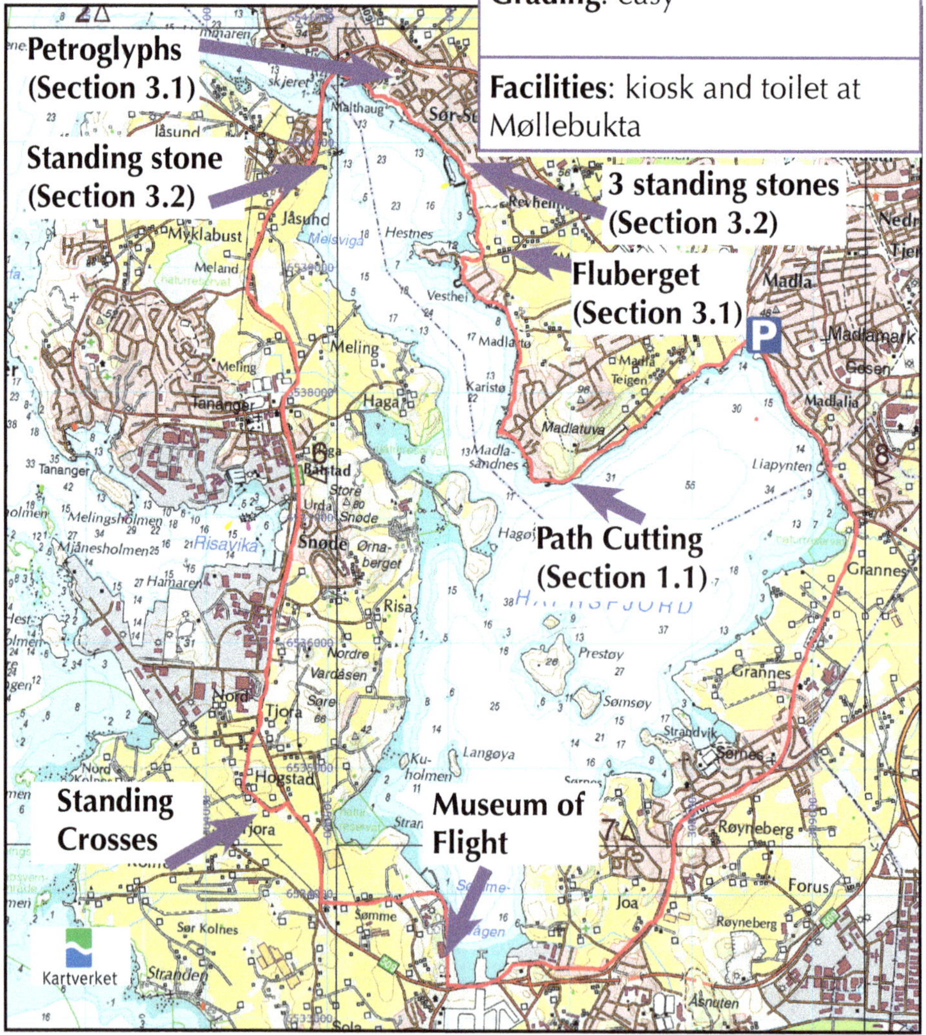

follow the signposted route along the shore of Madlasandnes to Hestnes. Just past Hestnes, the petroglyph site at Fluberget is about a 200m detour up the road. There are more ancient sites further along the shore: three standing stones, and further on, a little petroglyph site at Sør-Sunde.

The Hafrsfjord Bridge is past this site: leave the shoreline trail and join the cycle path accross the bridge. The standing stone at Jåsund is just on the far side, near the kiosk selling seafood. The cycle path is signposted beside the main road here. Follow it (watch for the point near the turnoff to Tananger harbour where it goes through an underpass to the other side of the road). Beyond Tananger, the site at Tjora is signposted along a small road to the right. After this, rejoin the main road/cycle track and follow the signs back to the fjord shore and past the Museum of Flight.

Watch for planes landing as you pass the end of the runway at Sola Airport, and then follow the signposted cycle route through Joa and along to Røyneberg (again, watch for where the route crosses the main road via an underpass). At Røyneberg, watch for the signposted route under another underpass to join the cycle route along the east side of Hafrsfjord. Cross the road just before the roundabout at Grannes to reach the final section of the route back to Møllebukta and the Three Swords.

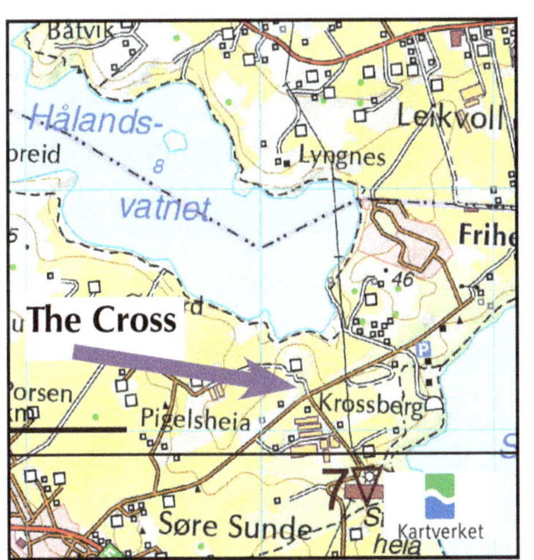

© Kartwerket

Excursion 3.3.2:

Visit the modern Cross at Krossberg

Although this site is beside a road, it is only a short detour from the popular trails around Store Stokkavatnet and Hålandsvatnet.

To get to the site at Krossberg:

By car: From Stavanger centre, take Rv509 towards Tananger. After about 6.6km, turn right at a roundabout along Krossbergveien. The site is on the left hand side of the road after about a kilometre. There is a little parking area.

Above: Obrestad Fyr, seen from Kongevegen

Chapter 4:
Getting around the Jaeren

Above: Hiking along Kongevegen

The rock carvings discussed in Section 3.1 give us a window into life in the Bronze Age. The petroglyphs left behind by the people who lived here show what can only be assumed must have been important to them. When it comes to transport, they carved only boats. Further to the east in Norway, rock carvings from the same age show carts as well as boats, but in all of the west of Norway, just one cart has been found so far.

Travel by boat was a necessity in those days: the interior of the country was largely impassable bog in the places where it wasn't slippery bare rock or impenetrable forest. It is not a coincidence that the early mediaeval kings of Norway chose to live at Utstein, in a very sheltered site on a tiny island. From here, the expert sailors of the time had easy access to all of Norway, around the coasts and along the fjords. With one exception, evidence of ancient routes across the land in this area is scant, and the few that there are tend to link fjords via short overland trails. The exception is the Jaeren coast.

This chapter looks at two of the older roads in Jaeren, how and why they were built, and where they may be seen today, and then visits a more modern coastal trail.

Above: The Jaeren shore, viewed from Kongevegen

4.1 Kongevegen
"The King's Road"

The Jaeren is unlike most of the rest of Norway. It is quite flat, and the area is relatively clear of the bare rock or forest that covers so much of the land elsewhere. Just 2.7% of the land area of Norway is farmable, and much of that is in the Jaeren. It is not that farming here has ever been easy though. The piles of stones beside many fields bear witness to centuries of toil: the stones had to be cleared away before the land could be ploughed for crops. The area was once a huge glacial outwash plain (see Chapter 2), and those stones were rounded as they were moved by the glaciers, and then deposited here near the ancient shore, giving a modern shoreline that is relatively straight and flat: a rather unusual occurrence in Norway.

Above: Looking towards Obrestad Havn from Kongevegen

This shoreline is dotted with ancient remains: burial mounds, boathouses, and habitations. There was a trail along the beach and pebble coast, along which people traveled freely. Traces of these old trails are rare to see nowadays. They are usually sunken tracks, or "hjulvegene" in Norwegian. A remnant of such a track may be seen at Grødaland. Folk history mentions a road along the Jaeren coast in Viking times, and that it was used by Olav Trygvesen as he traveled between family in Sola and Obrestad. At around AD 900, a law was passed concerning roads. It required that all roads were to continue to follow the routes they had followed since time immemorial, so it is likely that the route of this road is still used today.

Above: The old graveyard at Varhaug

But, the word "road" here is a bit misleading: we think of a "road" as a tarmac highway along which cars can speed without hindrance or punctures, whereas these "roads" were really trails. This "road" across Jaeren was at the time a path used for walking, or perhaps riding a horse or leading a packhorse along. There were no bridges, although there were some stepping stones that were regularly rendered useless when the water level was high. Nothing changed until the seventeenth century.

In 1636, local magistrates and sheriffs were told to allocate "road" maintenance to local inhabitants. Revisions of this law in 1643 and 1648 instructed local sheriffs to ensure that roads were built in their territory. This is probably what led to widening and improvements of the Jaeren coast trail, making it passable for carriages, at least in some places.

In 1704, King Frederik IV travelled across Jaeren, He rode his horse between Egersund and Ogna, but at Ogna Church he "met" with his chaise (a kind of carriage); we don't know how the chaise got there. The journey from Ogna to Sandnes took two days, and led to the renaming of the road as"Kongevegen", or "The King's Road". From Sandnes, rather than become mired in the bog between Sandnes and Stavanger, the King went by boat along Gandsfjord. Today, it takes about an hour to drive from Ogna to Sandnes along Rv44. From

Above: The church at Varhaug

there, the modern driver simply turns north on to the E39, and reaches the centre of Stavanger in less than half an hour.

Kongevegen was eventually replaced by the Rv44 slightly to the east, leaving the old road for the use of farm vehicles. In the early 1990s, Hå municipality negotiated with local landowners for access to create a hiking trail along the Kongevegen route. Today it is undergoing a renaissance: with several points of access, good waymarking, many interesting places to visit along the way, and stunning scenery, Kongevegen is popular with local hikers and tourists alike.

Below: One of the restored farm buildings at Grødaland. The farm is now a museum, located alongside Kongevegen

Excursions 4.1.1:
Hikes along Kongevegen

A long stretch of windswept coastline. Here there are fewer sand beaches, and more pebbles, with several interesting places to visit. As with the hikes further north in Chapter 2, hikes "there and back" are just as satisfying as those where transport is available at both ends of the section. It is possible to connect these hikes with the hikes in section 2.3 if desired.

Access	car
Distances	Saltebukta to Hå Gamle Prestegard: about 2km Hå Gamle Prestegard to Obrestad Fyr: about 1.3km Obrestad Fyr to Grødaland: about 4.8km Grødaland to Varhaug Old Church: about 3.7km Varhaug Gamle Kirkegård to Madlandshamma: 2.8km
Climb	negligible, but some short, steep slopes
Grading	easy
Facilities	Toilets, art gallery and cafe at Hå Gamle Prestegard, toilets and museum at Obrestad Fyr and Grødaland, toilet at Varhaug Gamle Kirkegard

To reach the start of the hike:
Car parking is available at Hå Gamle Prestegard, Obrestad Fyr, Grødaland and Varhaug Gamle Kirkegård.

Hå Gamle Prestegard: From Stavanger, take the E39 south towards Kristiansand, and leave after 12km, on Rv44 towards Bryne. Continue along Rv44 for about 25km, at which point turn right, signposted to Obrestad Fyr and Hå Gamle Prestegarad. Hå Gamle Prestegard is at the end of this road after 3.4km.

Above: Obrestad Harbour

Obrestad Fyr: From Stavanger, take the E39 south towards Kristiansand, and leave after 12km, on Rv44 towards Bryne. Continue along Rv44 for about 25km, at which point turn right, signposted to Obrestad Fyr and Hå Gamle Prestegarad. Obrestad Fyr is signposted to the left after 2.7km.

Grødaland: From Stavanger, take the E39 south towards Kristiansand, and leave after 12km, on Rv44 towards Bryne. Continue along Rv44 for about 27km, until you see "Grødaland" and "Kongevegen" signposted to the right. Turn here, and follow the track for about 500m to the car park on the left hand side of the road. **NB: from this car park, there is a delightful forest trail around a nature reserve. The nature reserve trail is signposted from one corner of the car park, but it doesn't lead to the Kongevegen trail on the coast. To reach the Kongevegen trail from the car park, return to the entrance, turn left and follow the signposts past the buildings of the farm museum to reach the coast and Kongevegen.**

Varhaug Gamle Prestegard: From Stavanger, take the E39 south towards Kristiansand, and leave after 12km, on Rv44 towards Bryne. Continue along Rv44 for about 30km. Turn right here, signposted Varhaug Gamle Kirkegard and Kongevegen, and follow the signs to the car park.

Below: An old boat winch rusts on the shoreline

Hike Notes

The trail is very well signposted along the length from Hå Gamleprestegard to Madlandshamma. South of Madlandshamma, Rv44 is close to the coast, and hiking becomes rather noisier.

Despite consisting of long stretches of relatively flat terrain, there is plenty here to observe and explore. Kongevegen is accessible all year round, making it a fine outing during the winter months when other areas are covered in snow. The area is worth many visits to do justice to all there is to enjoy here.

Excursion 4.1.2: Naerbø Museum

Access: car or train
Length: n/a
Climb: n/a
Grading: n/a
Facilities: toilets, cafe, picnic areas

The museum at Naerbø has the motto "learning by doing", and there is plenty here to amuse everyone. There is, as you would expect in Norway's premier farming area, a farm section with animals to meet, plus there is a science area with exhibits explaining the local agricultural science, landscape, climate, and more.

For current opening times and prices, see www.jaermuseet.no (this museum is the "Vitengarden").

To reach the museum:

By train: take the local train to Naerbø station. From the station, a short walking route to the museum is signposted.

By car: From Stavanger, take the E39 south towards Kristiansand, and leave after 12km, on Rv44 towards Bryne. Continue along Rv44 for about 26km, until you see a left turn signposted "Naerbø". Turn left along this road, and then turn right after about 1.5km. The car park is signposted on the left after about 800m.

Above: The museum at Naerbø

4.2 Building Roads: "The Western Highway"

Early road, or perhaps we should say trail, maintenance in the Jaeren consisted mostly of the kinds of techniques used today on popular hiking trails. Heather or brushings were placed on muddy sections, and possibly stone slabs or stepping stones were added in wet areas. This gave trails that, in summer, were passable for riders and packhorses. Rivers were crossed the hard way: by finding a place where the water was shallow enough to ford.

By the eighteenth century, Kongevegen, which had the potential for travel in a carriage or wagon for at least for some of the way, was the exception. Transport overland was by foot or horse, and most of the goods went by packhorse, both in Jaeren and beyond in Ryfylke. Laws were now in place that

Top: moorland seen from the "Western Highway"
Above: Sign at the south end of the "Western Highway"

made the local people responsible for road building and maintenance in their areas. Local sherrifs took on the organising and overseeing of the work. Several bridges were built, and the roads were generally improved. In 1745 it was noted that, in addition to Kongevegen, there was also a post road that ran south from Ålgård to Helleland, and from there to Egersund.

Above: Nordsjørittet 2016

Still, the roads remained in poor condition, and a governor called Scheel decided that there was an urgent need for a new road between Ogna and Egersund. Work began in 1789, and continued, with delays, for several decades. This "Western Highway" was not finally completed until 1843, but several parts of it were usable before then. The route followed the "Danish" tradition of building straight, up hill and down dale (one can but wonder if the Danes inherited this tradition in turn from the Romans, or if it was just easy to build straight roads over the flat Danish landscape!). Although the road was built for carriages and carts, in some places the gradient was so steep that there were problems for vehicles that were heavily laden.

At first, the sparkling new highway didn't make much difference to the local population. In 1805, Pram wrote that the new road "lies there with no wheel tracks and so little patronised that only occasionally does one meet an individual on horseback or a pack horse". We can surmise from auction records that nobody owned a cart, as none were traded in the area until 1822.

With time, and perhaps in part due to the steep gradients of the West Coast Highway, the main road from Stavanger to the south was routed further to the east. The West Coast Highway remains as a testament to the efforts of workers who built roads the hard way, by hand. Nowadays it is a pleasant cycle path or hiking trail, made all the more so by good train links that make it easy to traverse in one direction only. Every June, the "Nordsjørittet" cycle race takes this "highway" on a section of its route from Egersund to Sandnes. This cycle race also travels along some of the tracks and trails that remain of Kongevegen.

Excursion 4.2.1: Cycle the "Western Highway"

Easy access by train makes this cycle ride a delightful one-way trip from Egersund to Brusand.

Access: train
Length: 23km
Climb: around 200m
Grading: easy
Facilities: Town centre facilities 1km from the start at Egersund

To reach the start of the ride:
By train: Take a train to Egersund station. Current schedules and fares may be found at: www.nsb.no. Bikes are allowed on trains, but must be stored in designated places. It's worth checking before you travel, as sometimes space is limited. Most of the staff at the train station in Stavanger speak English, and will be happy to help.

Ride notes:
The route is reasonably well signposted with cycle route signs along the whole stretch from Egersund to Brusand.

From Egersund station, be sure to turn northwest (right) to begin the route. The first stretch follows an old railway track, complete with cuttings and one short tunnel. Beyond Helvik station, there is a short road section, first along a quiet road, then on Rv44 before the trail turns SHARP right up a short hill. At the end of this tarmac stretch, the section of the "Western Highway" begins. It ends just before Ogna, after which the route is well marked to Brusand. The beach there makes a lovely picnic spot.

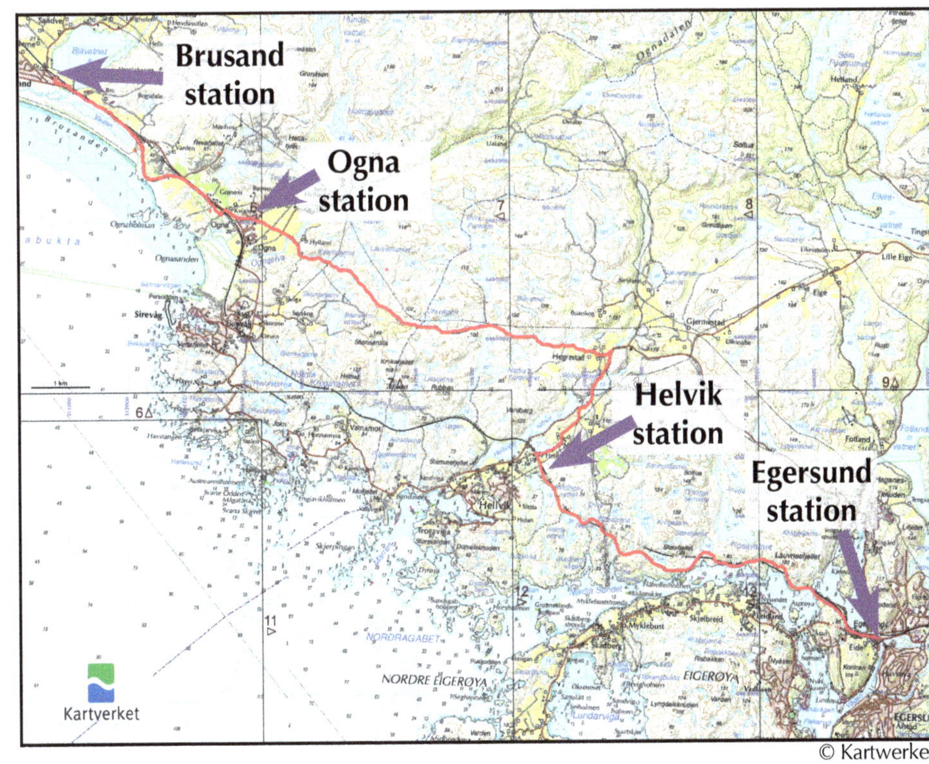

© Kartwerket

Excursion 4.2.2:
Hike the Old Road at Viglesdalen

Even without the old road, this valley is a splendid day hike or the beginning of a multi-day trip into the mountains

Access: car

Length: 13.3km

Climb: around 350m

Grading: moderate

Facilities: Toilet at the car park, cabin at Viglesdalen

To reach the start of the hike:

By car: From Stavanger, take the Tau ferry. After leaving the ferry, continue along Rv13 towards Hjemeland for about 23km. Here, take a sharp right turn along Rv661 towards Nes (if you reach the village of Årdal, you have gone too far). After 2km, carry on straight along Rv638, and then at the T-juction turn right towards Nes. The car park is at the end of the road, after about 4km.

Hike notes:

The trail is well signposted, and follows the old drove road to the summer grazing grounds at Viglesdalen. Before the road was built, local farmers were obliged to cross the lake at Vigdelsvatnet via a hazardous boat crossing. To make the journey safer, Swedish labourers were employed. They built the road between 1907 and 1912. The original plan was to build the road further and higher into the mountains, but by the time Viglesdalen was reached, most of the land beyond had been sold as hunting grounds, so the road building stopped there. From the cabin, either return the way you came, stay the night, or continue on a multi-day trip.

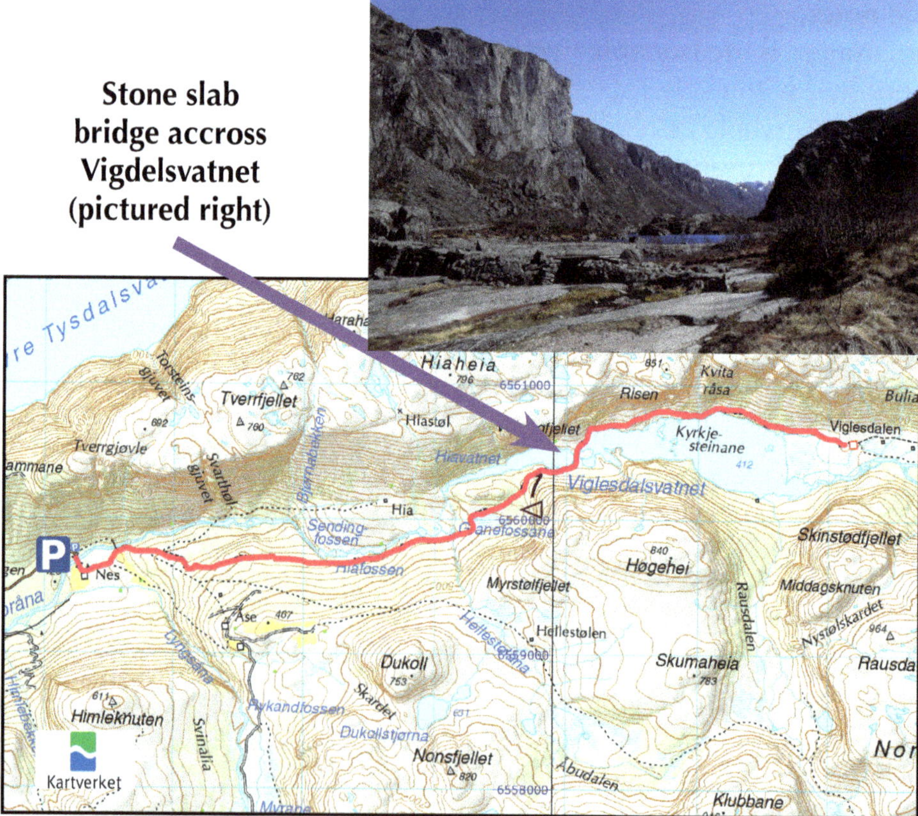

© Kartwerket

Excursion 4.2.3: **Visit Terland Klopp**

One of the earlier bridges to be built in this area, Terland Klopp today is more of an attraction than a working bridge.

Access: car
Length: n/a
Climb: n/a
Grading: n/a
Facilities: none, but road service area nearby on E39

To reach Terland Klopp:

By car: From Stavanger, take the E39 south towards Kristiansand. About 17km beyond Vikeså, turn left along Rv42 (signposted to Tonstad). The bridge is signposted on the right hand side of the road after about 6km. NB: travel time from the centre of Stavanger is about an hour.

Notes:

Terland Klopp is a stunning example of a stone slab bridge. It is 60 metres long, with a total of 21 arches. The width of the arches varies between one and two metres. The bridge that we see today is an improved and upgraded version of a bridge first built in the early 1800s. The pillars are built in drywall construction, and the arches are made from large slabs of stone laid accross them. The bridge continued to be used by local traffic crossing the River Giaåna until 1977.

The bridge was restored in 1986, and was officially protected as a cultural monument in 2008.

Above: Terland Klopp

Above: The rugged coast between Vigdel and Ølberg

4.3 The Wonderful, Crazy, Winding Path: from Vigdel to Rege

North beyond Hellestø beach, the coastline changes. Flat pebble shore gives way to small peninsulas and bays, with outcrops of sharp, jagged shale. Here, beyond the reach of Kings in their carriages, lies a different coastal trail.

Marked with blue markers and dots, it crosses beaches and climbs around rocky headlands. Along the way are tidal pools, wildlife, a restaurant, kiosk, and camping opportunities. The topography here is a little similar to the Kvernevik coast, but this trail feels a world away from the urban gravel trail of Kvernevik.

Top: Rock near Vigdel
Above: Ølberg beach

Excursion 4.3.1:
Hike From Vigdel to Rege

This delightful coast is worth returning to many times, to appreciate it in all four seasons.

Access: car

Length: about 4km in one direction, easy to shorten e.g. Vigdel to Ølberg is about 1.5km

Climb: negligible

Grading: easy, but some rocky sections

Facilities: Toilet at Vigdel car park, toilet, kiosk, restaurant, and campground at Ølberg, campground near Rege

To reach the start of the hike:

By car: From Stavanger, take the E39 south towards Kristiansand, and leave after 9.3km, signposted "Sola". After 5.3km, turn south (left) along Rv 510. Turn right along Nordsjøvegen after 2.5km, and then left after another 1.7km, signposted "Ølberg". At the T-junction after 2km turn left, and then right at the next T-junction. The car park is signposted on the right hand side of the road after another 150m.

Hike notes:

The route is well marked with blue posts and dots. Begin by turning north along Vigdel beach with the sea on your left, and look for the blue markers on the rocks beyond the sand.

The trail leads towards a rocky inlet: here a sign (in Norwegian) warns that the trail is unsafe when tides are unusually high or storms are raging. Please turn back and return another day if necessary!

From here, the trail is marked accross a field and around a little peninsula. It follows a rocky shore along the side of another field before arriving at Ølberg. To the left (north), there is a harbour,

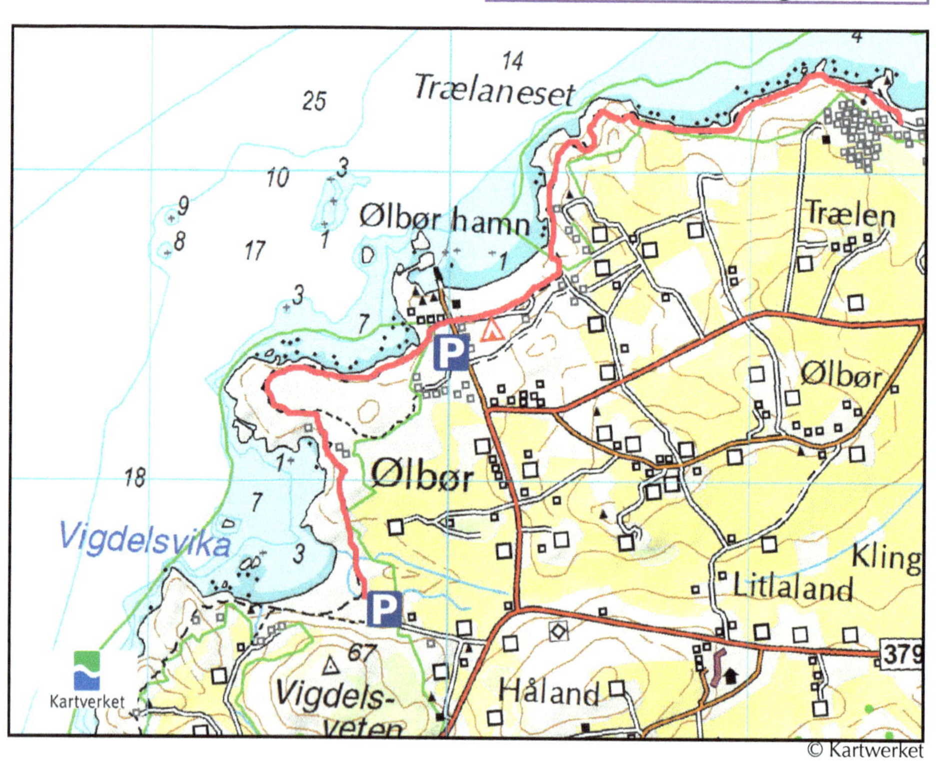

Above: A little pothole filled with ice near Rege beach.

whilst straight ahead is the campsite and kiosk. You can choose to walk along the campsite track, or turn left and then right just after the restaurant to cross Ølberg beach. Either way, rejoin the blue markers on the far side of the beach and climb away from the sand to enjoy views of Sola beach and Tananger.

Continue down the hill and along a section of rocky shore, where you may see shallow potholes (see section 2.4). Look to the right to see the sheds at Sola Allotment gardens ("Strandhage" in Norwegian). Notice how these are all the same, in contrast to the many different cabins at the allotment gardens (Kolonihage in Norwegian) in Stavanger. Rege beach is just beyond. It is possible to continue hiking to Sola beach, or turn around here and return the same way to Vigdel.

Excursion 4.3.2: **Frisbee Golf at Ølberg**

In 2015, a new 18 hole frisbee golf course opened. Bring a frisbee and stop on your way to or from the beach for ultimate frisbee fun!

To reach the course:

By car: From Stavanger, take the E39 south towards Kristiansand, and leave after 9.3km, signposted "Sola". After 5.3km, turn south (left) along Rv 510. Turn right along Nordsjøvegen after 2.5km, and then left after another 1.7km, signposted "Ølberg". The car park is on the right hand side of the road, immediately after the junction.

Follow the trail from the car park to the start of the course.

Note:

The course is operated by Sola Frisbee Klubb. for more information, see their website at www.solafrisbee.com.

Above: The frisbee golf course near Rege beach.

Above: Wooden houses at Gamle Stavanger

Chapter 5:
The Town of Wooden Houses

The east of Storhaug was once an area of canning factories and industrial bustle. With the demise of the canning industry, it declined. In the past few years the area has been rejuvenated: derelict factories have been restored as offices, trendy coffee shops have appeared, and new blocks of flats are going up, ready for modern city-dwellers to move in and enjoy the vibe. The old Tau brewery is now a thriving arts centre, and, just along the shore, is a very modern block of flats. Most blocks of flats are made of concrete, steel, bricks or stone. This block is made almost entirely of wood. This can't have been an easy feat: wooden beams don't lend themselves to large buildings.

Above and bottom left: the complex of flats at Siriskjeret is built on reclaimed land at the side of the shore.

This particular block of flats is indeed special. It is called Siriskjeret, and was built as part of a project launched in 2008 when Stavanger was the European City of Culture. The aim of the project was to commemorate the tradition of building here with wood. Stavanger has over 8,000 wooden houses, built in a huge variety of architectural styles over four centuries, with more being added every year. This chapter takes a look at some of them, in loose chronological order.

Left: Valbergtårnet

5.1 Valbergtårnet: The Watch Tower

Top: The cannons used to warn of fire at Valbergtårnet
Above: The view from beside Valbergtårnet over Byfjord

The centre of Stavanger is a labyrinth of streets circling a knoll in an area called Holmen. At the top of the knoll stands a stone tower, with a copper roof and views across all of the surroundings. Clearly it's a watch tower, and it is easy to imagine the early inhabitants of the town looking out across Byfjord and the surrounding countryside to check for approaching visitors, welcome or otherwise. But this is only part of the reason the tower is here.

There has been a tower on this hill from the earliest days of the city. The first houses here, as throughout the rest of Europe at the time, were built with wood. Heating and cooking were with open fires. Houses were built very close together, so even if you were careful, your house could be burned down owing to an accident several doors along the street. Local residents were required to possess two fire buckets per house, and when the duty watchers on the tower spotted a fire, they would raise the alarm. A runner was sent to the Cathedral, so that the bells could be rung. At this signal, everyone would bring their buckets to help fight the fire. Today's tower was built between 1850 and 1853, replacing several earlier versions. Watching for fire remained one of the duties of the watchmen right up until the last guard, Tobias Sandstøl, relinquished his duties in 1922. However, it is not clear how much help the new watch tower really was. In 1860, a huge blaze ripped through the city, and the brand new watch tower was the only building left standing on the hill.

Above: Fire is still hazard in Stavanger

Many fires have burned through Stavanger over the centuries, and this is one of the reasons why the earliest houses built here have not survived to the present day. These fires have left their imprint on the city, even though the buildings were rebuilt again every time. Walking along Østervåg, it is possible to see that the section of the street to the north is much wider than the southern part. This is because the fire of 1860 burned down everything up to the end of the wider section, whilst the houses in the narrower section were saved. When the street was rebuilt, the houses were set further apart. Looking down in Østervåg, it is possible to see that the date on the manhole covers is 1866. This is the date when the city houses were connected to mains water, giving them a better way to fight small fires before they took hold.

Fire is still a major concern for the city. Wooden houses in the heritage areas of the centre of town are required to install modern sprinkler systems. Fireworks are popular on New Year's Eve, but not in these historic areas!

5.2 Living the High Life, Merchant Style

The harbour side around Vågen is lined with restaurants, clubs, banks and shops. It bustles with people from early morning until late at night. The street along the west side of Vågen is called Strandkaien. Turning up a short, narrow alley from here leads to Nedre Strandgate, which is generally quiet. Alongside Nedre Strandgate is something unusual for the town centre: a large garden. Clearly it was (and still is) very well cared for. The slope has been terraced, and paths laid along the levels. Below the gardens is an old well, once the source of fresh water for the locals. At one time, this street was much busier.

In the middle ages, the area to the west of the town harbour at Vågen, was owned by the Cathedral, and was not built upon.

Top: Christmas lights on former warehouses at Vågen
Above: Terraced gardens along side Nedre Strandgata

After the reformation, in the 1680s, plots began to be sold there. By the end of the 18th century, the local economy was booming, and rich merchants had moved in. The area came to be called Stranden, or Straen in the local dialect. Houses and warehouses lined the harbour of Vågen (on both sides). The warehouses were built directly at the waterline, and featured large hoists on the upper stories that were used to winch merchandise to and from the ships. Some of these hoists are still visible today. The merchants themselves lived in grand houses behind their warehouses. A street, Nedre Strandgata, ran immediately alongside the houses. The terraced gardens across the street were at that time used by the residents.

Above: This warehouse near Badedammen is still directly next to the water, as it was when it was first built. This is how the warehouses around Vågen looked, until land was reclaimed in front of them.

This pattern was originally repeated on the other side of Vågen. Today there is also a wide area of reclaimed land in front of the remaining warehouses. The reclaimed land was created when sailing ships were replaced by steam ships. The steam ships needed deeper water in which to dock, whilst the warehouse owners needed the sparks from the steam engines to be a good distance away from the combustible wooden warehouses and their contents. Most of the warehouses in the centre of Stavanger are now fronted by a wide expanse of reclaimed land, but it is possible to see warehouses in their original position right by the water at some places in Storhaug, Engøy and Buøy.

Several of the merchant's houses are standing today. One of these is the Rosenkilde house. This house was owned by the family of that name, who were originally Danish but were in Stavanger by 1680. Peder Valentin Rosenkilde brought a house on Straen in 1796. He became a great man in the town, and eventually ripped down the house in order to built a new one. The house we see today was finished by 1813, at which time is was the second largest in the city, smaller only than the Kielland house at Ledaal. The house stayed with the Rosenkilde family until 1887, when the shipping crisis ravaged the local economy. Eventually, the house fell into the ownership of Stavanger Kommune, and it has been used for many purposes, including as a community kitchen. Today, it houses the Stavanger Chamber of Commerce, and other offices.

Above: Skagen 18

Excursion 5.2.1:
Enjoy Coffee at Skagen 18

Skagen 18 is one of Stavanger's iconic houses. Built in distinctive Rococo style, the oldest part of the house dates back to around 1700, whilst the facade is from 1787. The house is still in the hands of the merchant family who first built it, but today it houses a bustling coffee shop. Take a break from the shopping and enjoy the historical atmosphere!

To find Skagen 18:
On foot: From the Kulturhus(library) in the centre of Stavanger: stand outside the main entrance door facing into the square. Walk straight accross the square, and then turn right, leaving the square in the direction of the main harbour at Vågen. Cross straight over the next road that you meet, go down a short hill, and then turn right at a T-jnction. Skagen 18 is about 100m along on the left hand side of the road.

Excursion 5.2.2:
Visit the Maritime Museum

Below: An old wagon stored in the alley under the Merchant's House at the Maritime Museum

Located on the harbour at Vågen, the Maritime Museum is a gem. In addition to the shipping exhibitions, there is a restored warehouse with a sail loft as it would have been when in use. Also, one of Stavanger's original merchant's houses is preserved here, complete with opulent furniture, and a mirror used by the ladies to spy on people in the street.

To find the Maritime Museum:
The Maritime Museum is at Strandkaien 42, Stavanger. If you stand at the fish market in Vågen harbour looking towards Byfjord, it is on the left hand side of the harbour. For opening hours and prices, see their website at www.stavangermaritimemuseum.no.

Excursion 5.2.3:
Attend an INN event at the Rosenkilden House

Many INN events are held in the upper floor of the Rosenkilden house. On the way there, you will ascent the grand staircase and see some of the reception rooms that were designed to impress. The upper floor was originally the attic, with large wooden beams and beautiful round dormer windows.

Details of INN hosted events may be found on the Stavanger Chamber of Commerce website, www.rosenkilden.com

Below: The Rosenkilden House

5.3 The Country Residences

The urge to escape from the city to a bolthole in the nearby countryside is not a new one. Rich local businessmen in the eighteenth and nineteenth centuries built country houses where their families could enjoy peace and quiet. They tended also to have land attached, which the owners could then use to run a farm as a side business. In Stavanger, the most famous of these are Ledaal and Breidablikk, but there are several others, which at one time formed a ring around the city.

The first of the country retreats to be built was a house called Blidensol. It dates back to early in the eighteenth century, but may contain elements that have been recycled from earlier than this. The family who owned Bildensol were famous because of the artist Andrew Smith.

Top and above: Bildensol

Andrew Smith was a very talented woodcarver and businessman who moved to Stavanger from Scotland. He Norwegianised his name to Anders, and became known in history as the person who carved the spectacular Baroque pulpit now in Stavanger Cathedral.

Above: The Boathouse at Sølyst, with the mansion behind it

At first, Andrew Smith's descendants lived in the city and used Bildensol as a country cottage, but eventually they relocated there. With time, the land owned with the house at Bildensol was divided up into lots and sold for development. That development grew into Gamle Stavanger, the subject of the next section.

On the other side of town, a magistrate called Eiler Hagerup Schiøtz built his country retreat on a secluded island. The island was originally known as Little Hvidingsø, but after the house was built (between 1822-24), it became known as Sølyst. The mansion house can be seen across the water from many places in downtown Stavanger in the winter, when it isn't hidden by leaves. In 1860, Sølyst was sold to the Somme family, and in the following decade the house was expanded and upgraded. It was then one of the biggest houses in the area. The house and surrounding land remains in the same family today. So far, Sølyst has escaped the fate of dense housing development, and the little island is still mostly forest and parkland.

The same cannot be said of the Kohler house. This grand house is on the site of the former farm at Hillevåg. Today, the Rv44 passes right beside its doorstep, just to the south of the Hillevåg tunnel. It is called the Kohler house after one of the first people who owned it, but it was expanded by the person who owned it next, Wilhelm Hansen. Hansen was one of the richest business men in town, and he added an extra storey to the house, plus the verandah supported on posts. Hansen had come to the house by marrying the rich heiress of the Hillevåg estate, Frida. Frida became interested in tapestry weaving and found fame as an artist. It was a meeting of dynastic families until 1883, when the Hansen business empire went spectacularly bust.

Over the years, and through several changes in ownership, the Kohler

Above: The rear of the Kohler house is just beside Rv44 at Hillevåg.

house fell into disrepair, but recently it has been beautifully restored. Slightly behind the Kohler house is another house from the former farm. This house is preserved the way it was after an extension in 1854. It was moved a few metres away from its original position when the road tunnel between Rv44 and the E39 was built, and now houses the Frida Hansen museum.

Like Bildensol and most of the other "country residences", the land of the Hillevåg farm was eventually sold as lots for building land. As the city continued to grow, a very different kind of "country retreat" was created in Stavanger during the First World War. By now, the city contained many workers who had moved in from the country and knew how to farm, but owned no land. The city authorities decided to give citizens the opportunity to grow their own food in allotment gardens, or "Kolonihage"in Norwegian, created for the purpose. Four sites were created: two at Storhaug, one at Våland, and one at the edge of Eiganes.

Now it was possible to enjoy fresh air even if you weren't rich enough to own a country house. Anyone willing to grow things there and keep it tidy could rent a plot in one of the allotments. Towards the end of the 1960s, growing vegetables went out of fashion and there was a lull in demand for plots. Usage changed slightly: people began to grow bright flowers and use the gardens as much for recreation as food. Nowadays, plots are very difficult to rent, as few of the current tenants ever want to release one. The gardens are beautifully maintained, and open to visitors most Sundays in the summer (see Excursion 5.6, The Varden Tour).

Left: Allotment gardens at Rosendal and Ramsvik, Storhaug

Excursion 5.3.1:
The Buøy Tour

Access: car or bus
Length: 6.1km
Climb: negligible
Grading: easy
Facilities: none

Choose a calm day to enjoy this tour of the islands of Buøy, Engøy, and the island retreat of Sølyst. Besides lofty views from the two bridges, there are small beaches, harbours, parkland and warehouses to see.

To reach the start of the walk:
By car: From Stavanger, take the bridge towards Hundvåg. Immediately after the first bridge, turn left at the roundabout. There is a small car park about 200m along the road on the left hand side next to the recycling stations.
By Bus: the walk can be accessed from Sølyst or Engøy bus stops. For current schedules and prices, see www.kolumbus.no

Walk notes:
This is number 39 in the "Stavanger 52 Tur" series. A large scale map, and more information in Norwegian, is available for free download at www.stavanger.kommune.no.

The walk is reasonably well signposted with T-markers and "Buøyturen no. 38" stickers.

From the car park, look for the T-marker opposide the recycling station that shows the way through the trees to the viewpoint on the hill. From here, follow the markers around the tour. If you get lost, go back to the nearest bridge and rejoin the walk there.

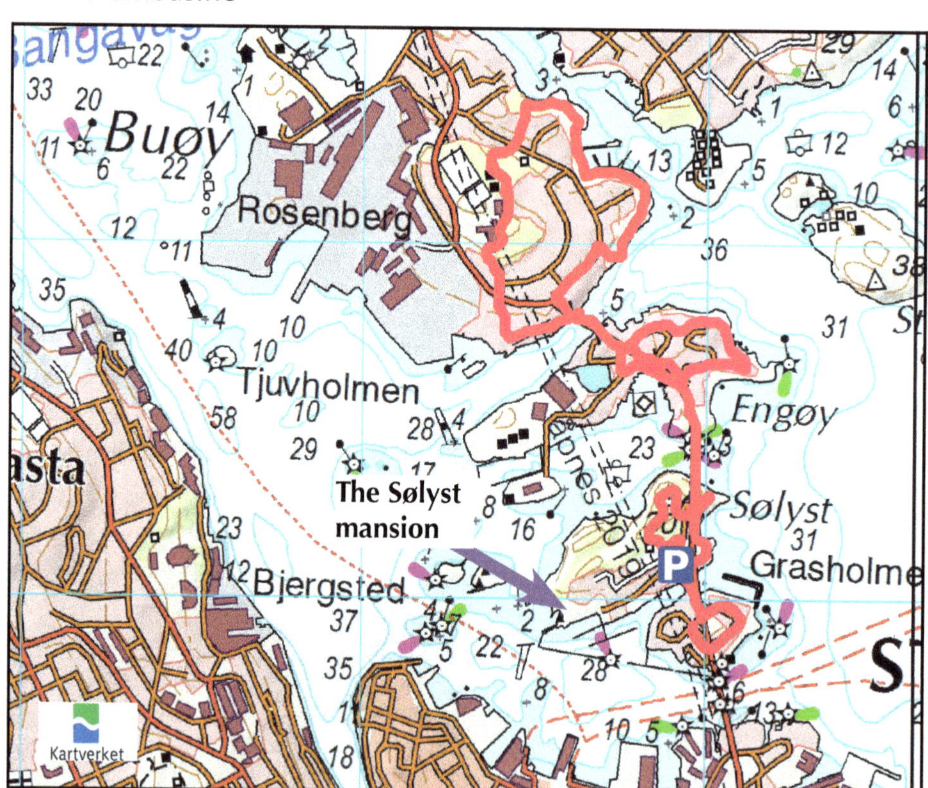

© Kartwerket

Excursion 5.3.2:
Eat at Charlottenlund

Charlottenlund is a beautiful house by the side of Lake Breiavatnet. Today, is it in the middle of the city centre, and it is strange to think that, when the house was first built, it was in the countryside. There is now a restaurant on the ground floor of the house, serving delicious finger food by the side of the lake.

To find Charlottenlund:
On foot: From the Kulturhus in the centre of Stavanger: stand outside the main entrance door facing into the square. Walk accross the square and exit along Laugmannsgata, towards the Cathedral. At the Cathedral, turn left and take the path to the lake (Breivatnet). Take the path along the side of the lake, with the water on your right hand side. Charlottenlund is on the left hand side, before the road in front of the bus station.

Above: Charlottenlund

Excursion 5.2.3:
Visit the Stavanger Museum

Below: Stavanger Museum

The Stavanger Museum contains a wealth of information about the growth of Stavanger, and how the people lived. At the time of writing, it was closed for refurbishment. Although it is not yet clear exactly what the new exhibits will show, the history of the city and the lives of its inhabitants is bound to feature prominently!

To find Stavanger Museum:
On foot: Stavanger Museum is at Musegate 16, 4010, Stavanger. From Stavanger Train station: Stand in front of the station facing lake Breiavatnet. Turn left, and then left again to walk up the hill (St Olav V's Gate). Cross straight over the roundabout by taking the underpass. Stavanger Museum is the white building on the left hand side of the road about 100m from the roundabout.

Above: Houses at Gamle Stavanger

5.4 Gamle Stavanger: The streets that were nearly demolished

Fire is not the only hazard that the wooden buildings of Stavanger have faced. Today, the area of Gamle Stavanger is a beautifully preserved melange of cottages dating back to the last part of the 18th century, but this area almost wasn't saved from the bulldozers.

The depression of the 1880s saw other shipping magnates besides the Rosenkilde family lose their properties along the Straen, and afterwards the canning industry moved in, slightly to the north. The grand houses and warehouses were joined by smaller houses, built for the factory workers in the canneries. In it's turn, the canning industry left the area. By the end of the second world war, Gamle Stavanger was almost a slum. The houses were crammed with people, who lived with unsanitary conditions, lice and poverty. After the war, town planners proposed that the whole

Top: A new use for an old stove
Above: Summer doorway in Øvre Strandgata

area be cleared, and replaced with modern buildings. One voice stood out.

Einar Heden was a local government employee who, in 1951, launched a plan to preserve a small area of these distinctive, dense wooden houses in the Straen area. At first, it wasn't a popular idea: few saw the value in this old, nearly derelict quarter. After a 5 year battle, the potential to renovate it was recognised, and Heden's plan was adopted by a majority of just one vote. The "Gamle Stavanger" (Old Stavanger) Society was formed in 1957, with Einar Heden in charge. It wasn't long before the society turned the tide of opinion. The preservation area was extended several times, without any battles, until by 1979 it included 170 buildings. Today, the area is beautifully kept, the residents are justifiably proud of their historic houses, and the area is a UNESCO World Heritage Site.

Above: A cobbled hillside alley. On one side, the cobbles are placed at an angle, to help stop horses and people from slipping on winter ice.

Most of the houses are traditional cottages, which would have begun as single storey dwellings. But, as more people moved in and needed space to sleep, families added a "half story" to the roof. A common modification was the "Stavanger dormer", created by lessening the pitch of the roof in the middle of the building to give an extra window. The attic room with the window was often reserved for guests, with the owners sleeping in the eaves at either side of it.

Left: Bergsmauet 2: This house was moved here from Sauda in 1831. Inside, it is possible to see the marks on the beams that were used to ensure that it was reassembled correctly in it's new home. Oral history tells that when people from Ryfylke emigrated, to the U.S.A., they sold their houses to pay for the transatlantic crossing. Some of the purchasers moved the houses to Stavanger.

Above: Christmas star

Excursion 5.4.1: **Visit the Christmas Market at Gamle Stavanger**

Large picture windows in the basements show that many of the houses in Øvre Strandgata were shops and/or workshops at one time. On the first Sunday in Advent, a Christmas Market is held. Both former shops and private houses open their doors, with gifts, decorations, and refreshments to sell.

To find the Christmas Market:
On foot: The Market extends throughout all of Gamle Stavanger. From the Tourist Information centre at the side of Vågen, follow the crowds up the hill into Gamle Stavanger. There is usually a friendly and festive atmosphere!

Excursion 5.4.2: **Visit the Canning Museum**

Below: The Canning Museum

Yes, there is a Canning Museum in Gamle Stavanger! It occupies a former canning factory, and gives a fascinating insight into the rise and fall of the canning industry in Stavanger. While there, don't miss the cafe in the beautifully restored worker's cottage next door.

To find The Canning Museum:
On foot: The Canning Museum is at Øvre Strandgate 88, Stavanger. From The Fish Market at Vågen: stand with your back to the water, and walk straight ahead up the road. Turn right past the bank, and then turn right again into Lars Hertevig's Gate. After about 100m, fork right into a path accross a garden. Øvre Strandgate begins at the end of the path, and the museum is about 200m along the street on the left.

Above: House near Pedersgata

5.5 Around Pedersgata

The beautiful area of Gamle Stavanger tends to get most of the glory and also most of the tourists. On the other side of the town centre, beyond the red brick St Petri Kirke (the Church of St Peter), is a similar area. There were factories here too, some of which are restored and now used as offices. The workers in these factories needed housing. Pedersgata is a long street that stretches beyond St Petri Kirke to Badedammen. On either side of it are sloping streets of worker's cottages.

This part of town did not have a conservation society looking after it, and, although most of it is beautifully restored now, it is a little "edgier" than it's sister across Vågen. Petersgata itself boasts a wonderfully eclectic selection of shops. Healthy lunch food is next door to pizza and kebabs, secondhand books rub

Top: Autumn colour
Above: Steps to the door in Pedersgata

shoulders with imported ornaments, and the Love Shop is difficult to miss. There a thriving art gallery, traditional florist, and nearby a "slow food" Italian restaurant that is so popular it's difficult to obtain a reservation. Here you will find some of the most delicious ethnic takeaway in town and possibly some of the least: the area has been a hub for the international inhabitants of Stavanger for a number of years.

Above: Many building walls in the Pedersgata area boast murals from Nuart festivals.

Above the steep slope of the hill by Pedersgata, the streets change. The houses are larger and further apart than the cottages on the hillside. They are mostly all similar, because these houses were built on a budget. Although they are quite large, the houses here were never villas. They were built to house more factory workers, and each one contained at least two and often several families. These are the "horizontally divided" houses that are still found in the residential property sale listings. Each family rented one floor of the house, perhaps getting either the attic or cellar too if they were lucky.

The streets here are named after places in the surrounding areas in the Rogaland region: Kvitsøygata, Taugata, Lysefjordgata, and so on. When people moved to Stavanger to get jobs in the factories, they tended to live together with others from the same place in the same street, so the streets were named after the places where the inhabitants came from.

Left: Larger, multi-occupancy houses higher on the hill

Above: Badedammen

Excursion 5.5.1:
Along Pedersgata to Badedammen

As the name suggests, Badedammen was once the only place in town where the factory workers could wash. Since then, plumbing and showers have reached the houses, but the lake is still a place to swim and relax, and Pedersgata is well worth exploring on the way.

To find the Pedersgata:
On foot: The road begins at the large red brick church, St Petri Kirke, in the centre of Stavanger. Follow it to the east, away from the church, until you have gone under the bridge to Hundvåg. Then, look for a left hand turn that will lead you to the lake at Badedammen. From Badedammen, either return the way you came, or go back via Verven and along the shore past the ferry docks at Fiskepiren.

Below: Absinthen

Excursion 5.5.2:
Sunday Grøt at Absinthen

Absinthen is a restored building very close to Badedammen. It is painted a very distinctive green, presumably due to its name. The building is let out as offices, and has a cafeteria. On the last Sunday afternoon of the month, the cafeteria serves grøt (a delicious kind of porridge), that most Norwegian of dishes. A truly local experience, which combines well with exploring Pedersgata and the area around Stavanger Øst.

To find Absinthen:
Absinthen is at Nedre Banegate 41, Stavanger. It is the distinctive green building. For more details of what is happening there, see their Facebook page, or www.absinthen.no.

Above: A Paradis villa

5.6 The Villas in "Paradise"

As Stavanger grew in the late nineteenth and early twentieth centuries, people making money began to look for plots of land for building. The purchasers were well off, and each could afford an architect to design their own individual, dream villa. The town expanded in a flurry of flamboyant houses. There were two main areas: Eiganes, and another smaller area to the southwest of Storhaug.

The north of Storhaug was industrial, crammed with noisy factories and worker's houses. The factories churned out unpleasant smells, making the area well worth escaping when possible. South, over the crest of the hill from the industry, there was a very different place. The land faces southwest, and slopes towards the sheltered inlet at Hillevåg. It was peaceful and quiet, and already locally known

Top: "Mock Tudor" elements
Above: "Jugend" style roof

as "Paradis" (Paradise) by the time people began to buy plots there. As the villas went up, the name for the area stuck.

There are "Swiss-style" houses, with overhanging eaves and ornate woodwork, "Jugend" villas, with their steep roofs and stylish doorways, and many variations on traditional and newer styles. Stavanger dormer windows (see Section 5.4) appear in steep "Jugend" roofs, and everywhere there are ornate doors and windows. Beside the original villas, there are now newer houses, built in functional modern styles, some from the second half of the twentieth century, and some very new "funkishuser", with their flat roofs and open, functional spaces. Whatever the shape of the house though, Paradis is still paradise.

Above: A former shop still has "picture" windows.

Left: "Swiss" Style

Excursion 5.6:
The Varden Tour

Access: on foot from Stavanger centre

Length: 7km

Climb: negligible: several short hills.

Grading: easy

Facilities: sea pool at Godalen, Sunday waffles at the Kolonihage, occasional ferry to Store Marøy from Breivig Harbour in the summer

This tour around the south of Storhaug takes in Paradis, the swimming area at Godalen, allotment gardens at Strømvig and Ramsvig & Rosendal, and a viewpoint at the top of Varden itself. The section of the trail between Godalen and Breivig Harbour was voted "Best in Stavanger" in 2015.

To reach the start of the walk:

On foot: From Stavanger Bus/Train station, stand facing Breivatnet lake. Turn right, and then right again along Kongsgata. Continue straight for about 350m, and then turn left into Kirkegårdsveien. Take the first right at Paradisveien, and join the walk here.

Walk notes:

This is number 10 in the "Stavanger 52 Tur" series. A large scale map, and more information in Norwegian, is available for free download at www.stavanger.kommune.no.

The walk is reasonably well signposted with T-markers and "Vardenturen no. 10" stickers.

The walk begins along the south shore of Storhaug, taking in allotment gardens (where waffles are served on Sundays in summer) and the bathing area at Godalen. It returns via a more urban route, but taking in two viewpoints at Varden and Paradis.

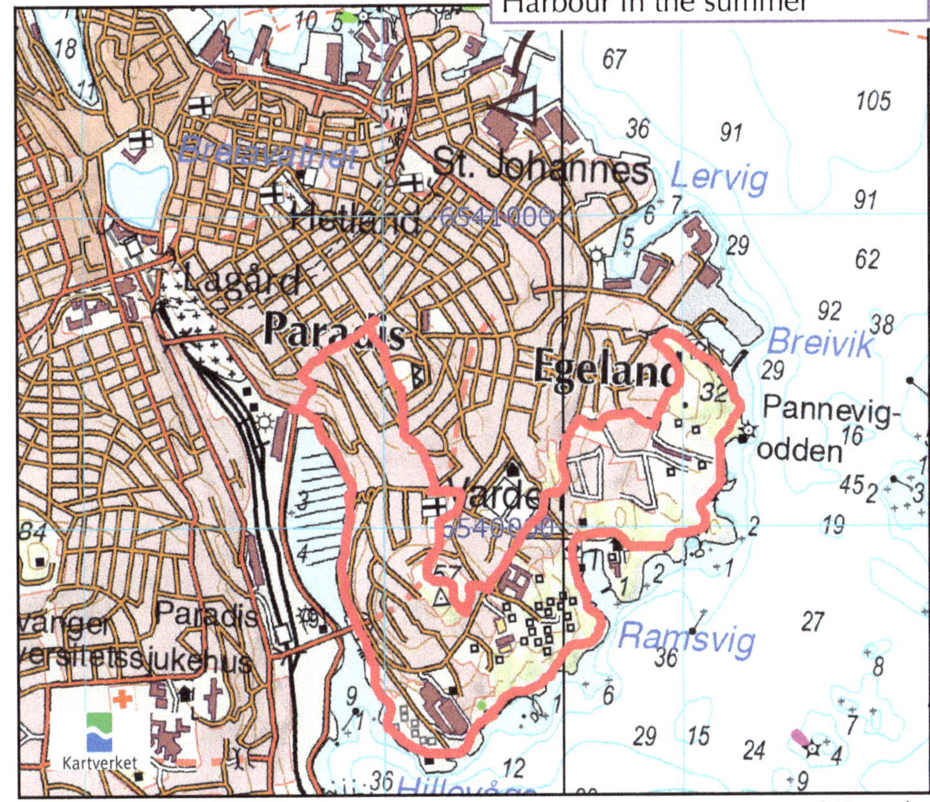

© Kartwerket

Above: Gingerbread shops

5.7 The Gingerbread Town

Gingerbread, or "pepperkake" in Norwegian, is an important feature of Christmas in Norway. This isn't just because it tastes delicious and children all over the country spend hours rolling dough, cutting shapes, decorating, and eating the results. Gingerbread is a building material. Gingerbread houses appear in many kitchens, nurseries and schools. Every year, Stavanger creates a Gingerbread Town. Iconic buildings from the city (and other places) are rendered in gingerbread slabs and sugar. They are joined by houses from stories, such as Hogwarts Castle, houses for Elves, towers, lights, and even vehicles. Every gingerbread creation donated to the town is displayed, with a model steam train chugging around the gingerbread town centre and Christmas music playing in the background.

Top and above: Gingerbread creations

The exhibition is open during

December, and is usually thronged with awestruck children and adults finding the magic hard to resist. For the past two years, the Gingerbread Town has been exhibited at the Stavanger Kulturhus. Don't miss it when it next appears.

Above: Gingerbread Tower

Left: Gingerbread Town

Clockwise from above:
Eiffel Tower, St Petri Kirke, Valbergtårnet, Ledaal

Above: The buildings in Øvre Holmegata were painted in bright colours to attract shoppers. Now the street is full of bustling cafes, and is locally known as "Fargegata": "The Coloured Street".

Chapter 6:
Reinventing Retail Through Changing Times

The founders of Amazon named their company well. Like the inexorable flow of a mighty river, the company, along with other internet retailers, has swept away a trail of retail outlets that perished before they could adapt to the new conditions. The large shopping malls that were the "new" innovation in the 1990s are now the "traditional" shops being rendered obsolete by the new technology. In Norway, the Post Office underwent a revolution. Post Office counters were set up in local supermarkets, operated by the supermarket staff. Gone were the restrictive opening hours: now everyone can fetch their packages at a convenient time. Internet shops can offer greater choice than retail, and usually compete on price. How do retail companies, shops, and shopping areas survive when the competition changes?

Above: Geoparken next to the Oil Museum. Fun places to play make trips to the city centre more attractive for families.

Left: Ut i Naturen sells kayaks from a restored warehouse by the waterfront. Here you can test your kayak before you buy it, unlike the landlocked discount warehouses or mail order suppliers.

Above: The Visitor Centre at the Figgjo factory

6.1 The River Figgjo: manufacturing quality tableware

To the south of Sandnes, the E39 crosses between Håbafjell and Bogafjell, then descends into the valley of the River Figgjo. Here, the towns of Figgjo and Ålgård were once small hamlets, but now they are both "dormitory" towns, where people live and commute to work elsewhere. A track along the side of the River Figgjo is heavily used by residents for recreation. Near the north end of the track, there is something unusual in the river. Broken plates, lots of them, are strewn along the riverbed. The plates aren't the leftovers from a recent Greek-Norwegian wedding; it's the factory at the end of the track that provides the explanation.

Top: Vintage Figgjo china designed for the catering industry
Above: China at the bottom of the River Figgjo

The Figgjo company was founded in 1941, and the factory provided china for the local market. By 1951, it was exporting chinaware to England and Denmark. The Figgjo company merged with its local competitor, Stavanger Flint, in 1968. In 1979 production was centralised at the Figgjo factory. Through the late 1980s and the 1990s, the company focused on good design, winning several awards. Retail times were tough though, as the market was flooded with cheap imports. In 1994, the company decided to pivot in a new direction.

Above: Figgjo children's china is available in local shops.

Hotels and catering businesses needed tableware that was robust enough to survive daily cycles in industrial dishwashers, yet still beautiful enough for guests to admire. The Figgjo china was well designed, and of excellent quality. The company decided to focus more on this market. Their sleek, modern china and quality were a huge success. Ever since the decision was made to focus away from the retail markets, Figgjo has been winning design awards and contracts to supply eateries as far afield as Dubai, Russia, London and Paris. If you take a peek underneath your cup or place in a Stavanger hotel, there is a high chance that the name "Figgjo" will be stamped proudly on the base. Figgjo haven't entirely abandoned their retail roots, though, and Figgjo children's designs can be found locally, in high end retail shops.

Excursion 6.1:
Visit the Figgjo Museum

At the Figgjo factory, there is a visitor centre with a cafe, outlet shop and museum. The museum tells the story of the factory and the company, with displays of their designs over the years. For opening times, see www.figgjo.no under "fabrikkutsalg".

To find the Figgjo Museum:
By car: From Stavanger, take the E39 south towards Kristiansand. After about 22.5km, exit right on Fv220, signposted Figgjo. The factory and visitor centre is immediately to the right after crossing the river. There is a car park outside the visitor centre.

6.2 Fretex: upcycling history

Fretex is part of the Norwegian Salvation Army. They run a chain of thrift shops across the country, raising funds to help pay for the community services that they offer. The shops give more than just the chance to snag a barely-used bargain dress (although there is this possibility too). One corner of every shop is dedicated to upcycling. Hardback covers from old, damaged books are turned into retro covers for new notebooks. The unworn parts of worn woollen blankets become hot water bottle covers and bags. Even old cotton sheets are cut into "filler": cloth squares to use for cleaning, polishing, and anywhere else where we could be re-using fabric instead of allowing new polyester cloths to shed micro-fibres into the ocean.

Modern Scandinavian living trends have given the Fretex shops other treasures to be re-

Top: Vintage china at Fretex
Above: Upcycled products at the Madla branch of Fretex

homed intact. Weaving is a Norwegian tradition, and at one time, many houses would have had a loom for making rugs, blankets and tablecloths. The looms can be big, and when a house runs out of space, the loom sometimes comes to Fretex to look for another home. The same applies to spinning wheels, mangles, woven carpets and even items decorated with Rosemaling, a traditional Norwegian craft. Designs, often flowers, are handpainted in a distinctive style on to wooden plates, chests, and other items. A trip to a Fretex shop is often a showcase for local crafts.

Above: Rosemaling

Below: Rosemaling, tapestry, painting, and books

Fretex also collect locally made china. If you loved a discontinued line in the Figgjo museum in the previous section, this is the place to find it. There may also be china from Egersund, Stavanger Flint and other Norwegian manufacturers. Don't be too surprised if your Figgjo china has been donated to Fretex by a local business, and has words upon it such as "Grand Hotel"!

Fretex are justifiably proud of their treasures, and recently began a different method of selling. They now offer "cultural evenings" to groups, where the provenance and history of the items is described. It's interesting and fun to learn with friends, and perhaps take home a piece of history.

Excursion 6.2:
Visit a Fretex Shop

There are Fretes shops at:

Madla Amfi Shopping Centre
Langflåtveien 15, Marierio, Stavanger
Breigata 20, Stavanger Centre
Gjesdalbakken 3, Sandnes
Torneroseveien 7, Lura, Sandnes

Each shop has its own character, although those with more floor space tend to have more furniture for sale.

6.3 Creating Fun: Kvadrat and the city centres

Kvadrat is the local mega-mall, with one of its slogans being "everything under one roof". It's certainly true that most of the Norwegian retail brands are there, and you can get from one to another without getting wet.

The place had been added on to over the years, and is now a bit of a labyrinth. Getting repeatedly lost is fun for some although not for others. Eventually, most people learn their way around, but the crowded, narrow corridors, low ceilings, and dark, impossibly busy coffee shops weren't most people's idea of fun. Shopping on the internet in the comfort of the sofa beats the Saturday crowds for many.

If people are going to choose to go shopping when they can get the same products, perhaps for less, online, then visiting the mega-mall has to be enjoyable,

Top: The new play park outside Kvadrat
Above: The new atrium at Kvadrat, with high ceilings and natural light

and Kvadrat has just had a makeover. Outside the entrance, a flower shop with lots of pots for people with shopping bags to trip over has been replaced with an open expanse and a small playpark where kids can let off steam. A new facade has been built. Where before, the entrance door led to a little atrium with a low ceiling, now the atrium ceiling is high and has been built to allow as much natural light as possible. New coffee shops and fast food outlets have been added; instead of filling the area with tables for people to have to go through or around, the seating is cleverly tucked away in corners. The first impression has changed from "crowded and dark" to "light, open and relaxed". It was a little change, but it has made a big difference.

Above: Cafe tables in Øvre Holmegata

The centres of Sandnes and Stavanger have been improving too. As shops began disappearing, the local business associations began to formulate strategies for enticing people to visit. Sandnes has built a large wooden pagoda, called "Lanterna" at one end of the main shopping street. It forms a stage for entertainment acts, activities and information during the various events that are held in the town. Christmas is a special time for shopping, and "Lanterna" provides the focus for carol singing, glogg, pony rides and Christmas Elves. The shops participate in the festivities: making it fun for people to come into town means that they will shop here too.

Stavanger centre also makes the most of its pleasant streets, many of which are pedestrian only. Several years ago, a shop owner in a street called Øvre Holmegata suggested that the buildings in the street be painted in bright colours, to make it a fun place to visit. Now, the street is lined with cafes and thronged with tourists taking holiday snaps. The name of the street is changing, too. If you mention "Fargegata" (The Coloured Street) to any local, they will know exactly which street you mean. Play parks and street markets now complement the regular festivals held in town.

In December, a huge Christmas tree appears in the market square outside the Cathedral, and an annual "living advent calendar" hosts live events all around the town, one for every day in December up until Christmas Eve. Christmas is also a special time along the Jaeren coast in the town of Egersund. Here there is an annual Christmas market. What it doesn't have in size, it makes up for in character. There are stalls, seasonal food and a big (heated) tent with a programme of entertainment every day. Egersund Church, a beautiful and historic building, opens its doors offering peace and quiet to anyone who would like a break from the bustle. The market has become so popular that extra trains are laid on for people to travel there from Stavanger and Sandnes. The internet doesn't have all the shoppers yet.

Above: A stall at the Egersund Christmas Market

Below: The doors of Egersund Church are open during the Christmas Market

Excursion 6.3:
Visit the Egersund Christmas Market

The Egersund Christmas Market is usually held during the first two weekends in December, with an evening program on Thursdays and Fridays, and a daytime program on Saturdays and Sundays. For program details and opening times, see www.julebyen.no

To get to Egersund Christmas Market:

By train: take any train to Egersund Station. A free bus transport service is provided to the market, or it is a 10-15 walk from the train station.

By car: From Stavanger, take the E39 south towards Kristiansand. After about 64km, turn right along Rv42, signposted Egersund. After about 9km, you will enter Egersund. Follow directions to the (free) car park.

Above: A mediaeval carving found near the foot of a pillar near the north door of Stavanger Cathedral. It is believed to represent the hands of the Lord holding back the evil.

Below: This carved face is thought to have been part of the Cathedral choir that was burnt down in 1260. It is now in the Archaeologica museum.

Further Reading

These references aren't mentioned specifically in the text, but provide interesting and/or useful information for explorers.

In English:

"From Beaches to Mountains: 38 walks and hikes in the Stavanger Region", Ute Koninx and Rosslyn Nicholson, 2016, ISBN: 978-82-303-3372-3: This is a guide to day hikes in the area, with general information about weather, equipment, and navigation in addition to detailed guides for specific walks and hikes.

The first footsteps: Stone Age Trail through Rogaland, AmS Småtrykk number 67: Information leaflet about the Stone Age sites in Rogaland from the Archaeological Museun.

Mediaeval Monuments in Rogaland, AmS Småtrykk number 74: Information about mediaeval sites in Rogaland from the Archaeological Museum

"Will and Vision, a Pictorial Tale of the Stavanger Region", Hans Eyvind Naess, 2007, ISBN: 978-82-993652-7-7: a lavish and fascinating book with photographs and local history. The text is in both Norwegian and English.

"Rogaland's Beautiful Buildings in Wood: Culture and Tradition", by Hans Eyvind Naess (English translation: Rolf E Gooderham), ISBN: 82-993652-6-0: contains descriptions of interesting wooden buildings all around Rogaland

"Eloquent Vandals: A History of Nuart Norway", Kultur Forlag, 2011: the story of Nuart, from the beginnings of the festival in 2001.

In Norwegian:

"Fra tråkk til motorveg", Sven Magne Olsen, Statens Vegvesen, Rogaland 1995: A history of Norwegian roads.

"Øks et Aks, Helleristninger og Hus fra yngre Bronsealder i Rogaland", AmS Småtrykk number 40: Early Bronze Age sites in Rogaland

"Lokalhistorik Årbok for Rogaland", Rogaland Historielag 2016, ISBN: 978-82-90087-82-6

The Norwegian state directory of monuments online:
www.fornminner.no

The Sola Kommune page about the Stone Circle and its history:
http://domsteinane.sola.kommune.no/index.html

History of Paradis and Kolonihage:
www.erlingjensen.net

Other sources of information:

Stavanger Library is located at the Kulturhus in the centre of the town. It is very easy to join, and has a good English language section as well as audiobooks in English to rent. The ground floor has periodicals in the reading area. This is a very good place to ask questions!

INN Stavanger is part of the Stavanger Chamber of Commerce. They offer many trips and activities in English. For more information, see their website at www.rosenkilden.com

PWC (People Who Connect) is a social group offering a range of activities, including a Hiking Group and an Explorers Group. For more information, see their website at www.pwc-stavanger.no

Below: Oanes seen from the summit of Uburen

Below: Daffodils, Øvre Holmegata

Acknowledgments

Above and below: Street art near Pedersgata

This book began with a small idea, and came to occupy large amounts of my time during my last few months in Stavanger. As with most projects, I encountered several dead ends and changed my plans along the way. I could never have produced anything without the support and help of many friends.

Ute Koninx listened without laughing, encouraged me, and gave me invaluable feedback (not to mention proofreading). Ulrike Weiss gave me a "tour" of the city, and showed me several things I hadn't seen before. Wendy and Tor Minsaas showed me Solbakk and Gloppedalslura, amongst many other places over the years.

I have been fortunate enough to enjoy wonderful company while out exploring: thank you to the people of the PWC hiking group for many happy outings. Thank you also to the people who came along to the ISS Parent Walks in 2014-15 and the INN Active "Walks with friends", who listened patiently while I enthused about things to see on the walks.

Hugh and Hazel Nicholson helped me improve my ideas, talked over plans with me, and laughed when the printer ran out of map-coloured ink.

This project is much better for all of your help. Thank you.

Rosslyn Nicholson

Stavanger, May 2017

www.ingramcontent.com/pod-product-compliance
Ingram Content Group UK Ltd.
Pitfield, Milton Keynes, MK11 3LW, UK
UKHW050136280726
14058UKWH00006B/668

9 788230 335819